WOOD PRESERVATION

A guide to the meaning of terms

THE RENTOKIL LIBRARY

The Woodworm Problem
The Dry Rot Problem
Household Insect Pests
Pests of Stored Products
The Conservation of Building Timbers
The Insect Factor in Wood Decay
The Cockroach (Vol. I)
Termites—A World Problem

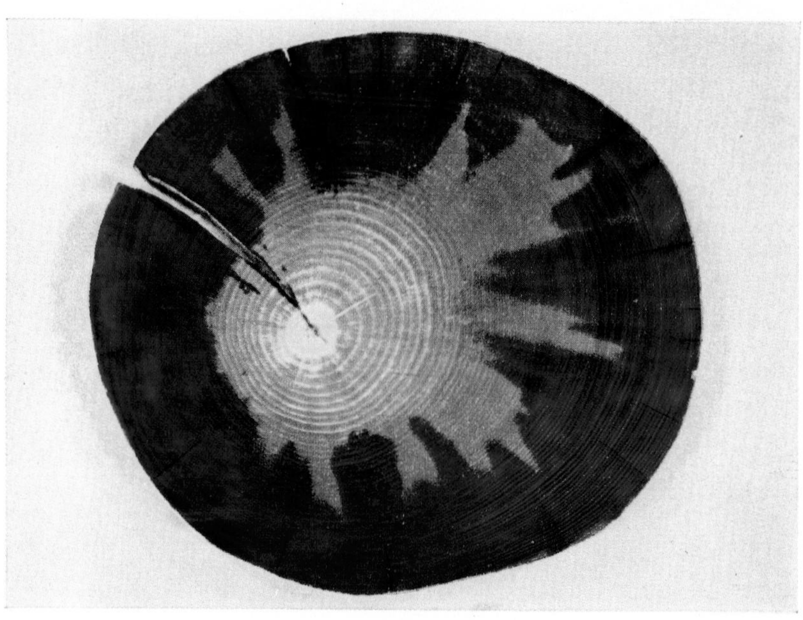

Transverse section of vacuum-pressure treated *Pinus ponderosa* using C.C.A. preservative. *Below:* Chrome azurol S reagent has been used as a stain which produces a blue colour in the presence of many metal cations (including copper), and red on unpreserved timbers. Differences in penetration, due to cracks and degree of seasoning are shown. *Above:* Unstained, for comparison. Boreholes show where timber has been extracted for chemical analysis.

WOOD PRESERVATION

A guide to the meaning of terms

NORMAN E. HICKIN

Scientific Director
Rentokil Laboratories Limited

HUTCHINSON OF LONDON

HUTCHINSON & CO (*Publishers*) LTD
3 Fitzroy Square, London W1

London Melbourne Sydney Auckland
Wellington Johannesburg Cape Town
and agencies throughout the world

First published 1971

This book has been set in Times type, printed in Great Britain
on art paper by Anchor Press, and
bound by Wm. Brendon, both of Tiptree, Essex
ISBN 09 101880 3

CONTENTS

ACKNOWLEDGEMENTS

The compilation of the definitions and descriptive texts contained in this work has not been accomplished without a great deal of help from a number of organisations and from many of my friends and colleagues. Firstly, I would like to acknowledge the help given to me by the British Crop Protection Council in allowing the use of text matter from the very useful Pesticide Manual published by them. An extensive critical examination has been made of several glossaries, definitions and standard terms published by the British Standards Institute, the British Wood Preserving Association, and the American Wood Preserving Association. Amongst my colleagues, Dr. David Dickinson gave considerable help with the mycological terms, Robin Edwards with those entomological, and Ian Stalker with chemical and process information. Vernon Hancock gave much assistance with building terms and John Evans with words connected with forestry. I am deeply grateful to these gentlemen, and I hope they feel that their efforts have helped to produce an instrument to further the technical competence of their colleagues.

INTRODUCTION

The object in producing this guide to the meaning of words used in wood preservation literature is to provide a more detailed and comprehensive work for reference purposes on this subject than exists at present. *The Glossary of Terms relating to Timber Preservation* published by the British Wood Preserving Association in 1962 has served its purpose well. The last few years, however, have seen a great expansion in several fields, such as in the remedial treatment section of the industry, with its own technology and expertise and its wide use of terms commonly used in building construction and maintenance. Terms now in general use in this important aspect of wood preservation, including those used in surveying, identification of wood-destroying organisms, and building construction, are therefore listed. The last few years have also seen the provision of standards both of the British Wood Preserving Association and the British Standards Institute containing much chemical and analytical information. The opportunity has been taken to incorporate many terms in this field into the list. In addition, the toxicology of wood preservative chemicals has, in recent years, been studied to an increasing extent, and as this is a field with its own special vocabulary, a number of toxicological terms has been given.

In deciding on the form which this guide was to take, much importance was attached to the speed at which a reference could be obtained. It will be seen, therefore, that the whole guide is in alphabetical order, not divided into sections but with all listed words categorised under one (or more) of the abbreviations listed below. This method defines the particular discipline involved, although there is some overlapping.

bac Bacterial, appertaining to bacteria.
bio Biological, appertaining to living organisms.
bot Botanical, appertaining to botany.
bldg Building, appertaining to buildings.
chem Chemical, appertaining to chemistry.
ent Entomological, appertaining to insects.
equip Equipment.
for Forestry.
myc Mycological, appertaining to fungi.
name Name of company, organisation, association, or proprietary product.
phy Physical, appertaining to physics.

proc	Process or method.
tox	Toxicological, appertaining to toxicology.
zoo	Zoological, appertaining to animals.

There is much variation in the amount of space given to individual words. In studying the explanations and meanings of words given in a number of standard works it was often found that additional words had been inserted with the idea of imparting more precision to a definition but which had the opposite effect, giving ambiguity or some other sense. So that in this guide some meanings of words are expressed in few words, but for some of the newer chemical substances it was thought desirable to add information not generally available in a single work concerning wood preservation.

Many words possess a number of meanings, but only that specifically applicable to wood preservation has been included.

The nomenclature of timber in general use in Britain has, to say the least, given many difficulties in the past, but the opportunity has been taken in this guide to follow the excellent example given in the British Standard Nomenclature of Commercial Timbers BS881 and 589: 1955, where the popular or common names of a timber are related to the botanical genus. Where there is no common name then the standard name adopted is the botanical generic name.

Some words are best forgotten. They never meant what they had every appearance of meaning, but, with time, the meaning of words changes and this is true in the wood preservation industry. New words are coined to express new ideas, and as we develop towards greater precision in our work, so words are found to express these new standards of exactitude and to give a sharper cutting edge to the words we use in this industry.

A

Abies. (*for*) A genus of north temperate softwood trees the standard name for whose timbers is fir, with the exception that *Abies alba* and *Abies pectinata* from central and southern Europe form part of what is known as Whitewood, but also known as Silver Fir. *Abies amabilis* from western Canada and western USA is Amabilis Fir. *Abies balsamea* from North America is Balsam Fir. *Abies grandis* from British Columbia and western USA is Grand Fir. *Abies basiocarpa* from British Columbia, Alberta, and western USA is Alpine Fir, and *Abies procera* (formerly *Abies nobilis*) from western USA is Noble Fir. Several species of *Abies* have been grown for some years in the British Isles and small quantities of such timber are marketed. It should be noted that the word fir is in use for a few species of softwoods other than of the genus *Abies*. See *Pseudotsuga taxifolia* and *Pinus sylvestris*. The wood of the true firs is nearly white with yellowish streaks. It is soft, light, strong, and elastic, and not very resinous, but is generally not such good timber as spruce, which it resembles. Small hard knots are usually present. It is in common use for medium-class joinery but cannot be used out-of-doors without preservative treatment.

Absorption. (*phy*) See Initial Absorption. Of gases, a solution of a gas in a liquid and sometimes in a solid. Of heat and light waves, that part neither reflected nor transmitted by the surface on which the waves fall and usually accompanied by a rise in temperature. The passage of a nutrient solution through the wall of a living cell.

Accelerated test. (*proc*) A laboratory or field test in which the conditions are arranged to simulate in a short time the effects of a more extensive period of natural ageing. This is often achieved by using wooden samples which are less than the normal size, containing low net retentions of preservatives. In the laboratory, intensive cycling of temperature, humidity and ultraviolet irradiation may be involved.

Acer. (*for*) A genus of north temperate hardwood trees: a number of species are of economic importance and usually referred to as maple, except *Acer pseudoplatanus*, known as sycamore in Britain.

Activation. (*chem*) Equivalent to synergism, *q.v.*

Acute dose. (*tox*) An amount of a substance which when administered causes a temporary but severe harmful effect.

Adsorption. (*phy*) See Sorption. Usually considered as a condensate of the molecules of a gas as a film on the surface of a solid, or of solute molecules at a solid/solution interface.

Adult. (*ent*) In insects, the final or reproductive stage of development, previously known as the *imago*, pl. *imagines.*

Adze. (*equip*) A long-handled tool like an axe but with the cutting edge at right-angles to the handle, used for forming a plane surface on round timber. Now obsolescent, but beams of many old buildings show adze chop marks which are sometimes reproduced in order to give an antique effect.

Aerosol. (*equip*) Fine particles of a liquid suspended in a gas, e.g. air as fog or mist. Now commonly used to designate an insecticidal chemical dissolved in a liquefied gas which is kept liquid under pressure in an appropriate container which may be a tin of the aerosol 'bomb' type, or of a more substantial re-usable type. A common 'propellent' in which insecticides are dissolved for this purpose is the refrigerant gas dichlorodifluoromethane. This substance possesses a vapour pressure of about 75 lb per square inch at 20°C.

Agar. (*myc*) The dried hydrophilic colloidal substances manufactured from certain seaweeds, used in the preparation of nutrient media for growing micro-organisms. The nutrients present determine the name of the medium, e.g. Malt Agar*, q.v.*

Agar-block test. (*myc*) A technique for evaluating the fungicidal properties of wood preservatives similar in many respects to the solid-block method, but nutrient agar takes the place of soil.

Agar-plate test. (*myc*) A test for preliminary screening of the fungicidal properties of a preservative in which the latter is dispersed at various dilutions in malt agar medium and inoculated with the test fungus. See *Malt agar*.

Agathis. (*for*) A genus of southern hemisphere softwood trees known as kauri. *Agathis alba* from the East Indies is East Indian Kauri: *Agathis australis* from New Zealand is New Zealand Kauri (this has been known for many years in Britain as kauri pine). Three species from Australia are known as Queensland Kauri and *Agathis vitiensis* from Fiji is known as Fijian Kauri. The relatively durable timber is light yellowish brown with even grain and good strength and elasticity properties. It has the property, rare in timber, of shrinking longitudinally. It has been used extensively for pattern making, boat decking and other purposes where some of the properties of a hardwood are required with easy working.

Air-dried. (*proc*) The condition of timber that has been dried by being stored in the open air, with or without overhead protection as distinct from drying in a closed chamber (kiln-drying). The condition of timber, the moisture content of which is approximately in equilibrium with local atmospheric conditions. In Great Britain the moisture content of air-dry timber may range from 14 to 23 per cent, according to season and timber species.

Alder, common. (*for*) See *Alnus*.

Alder, grey. (*for*) See *Alnus*.

Aldrin. (*chem*) $C_{12}H_8Cl_6$ (365) The common name approved by ISO, except in Canada, Denmark and USSR, and by BSI for a material containing not less than 95 per cent of 1,2,3,4,10,10-hexachloro-1, 4,4a,5,8,8a-hexahydro-exo-1,4-endo-5,8-dimethanonaphthalene. In Canada, aldrin refers to the pure compound, known as HHDN in Great Britain. It was introduced, in 1938, by J. Hyman & Co. as 'Compound 118' under the trade name 'Octalene' and its insecticidal action was first described by Kearns, C. W. *et al.*, *J. econ. Ent.*, 1949, **42**, 127; protected by USP 2,635,977.

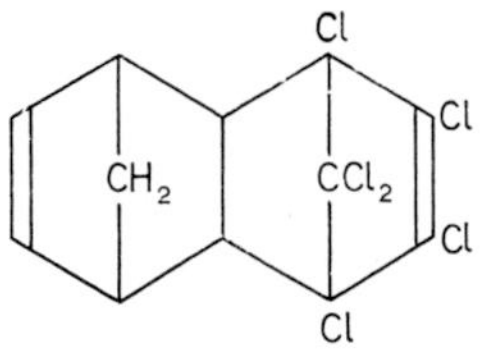

Cl
Cl
CH_2 CCl_2
Cl
Cl

The dehydrochlorination of the Diels-Alder adduct of cyclopentadiene and vinyl chloride yields bicyclo(2,2,1)-2,5-heptadiene which is condensed with hexachlorocyclopentadiene. The resultant compound is a white crystalline, odourless solid, mp 104 to 104·5°C, vp $2{\cdot}31 \times 10^{-5}$ mm Hg at 20°C. Though practically insoluble in water, it is moderately soluble in petroleum oils, readily soluble in acetone, benzene, xylene. The technical product is a tan to dark-brown solid, melting range 49 to 60°C.

Aldrin is stable to heat, to alkali and mild acids, but oxidising agents and strong acids attack the unchlorinated ring. It is compatible with most pesticides and fertilisers but is corrosive because of the slow formation of HCl on storage. It is a non-systemic and persistent insecticide, effective against soil insects at rates of $\frac{1}{2}$ to 5 lb/acre and is non-phytotoxic. It has been widely used in USA as a soil treatment for pre-construction use against termites. The acute oral LD50 for rats is 67 mm/kg; it is absorbed through the skin. Rats fed for two years at dietary level of 5 ppm suffered no ill effects but liver damage resulted at the 25 ppm level.

For the analysis of aldrin and its formulations, total chlorine is determined by sodium biphenyl reduction.

Residues may be estimated by reacting with phenyl azide, coupling the product with diazotised 2,4-dinitroaniline and measuring at 515 mμ, or by gas chromatography.

Alga. (*bot*) pl. ae. A member of a major group of the *Thallophyta.* The Algae are distinguished by the possession of special pigment within the cell or cells. One such pigment is chlorophyll. They are capable of assimilating and providing independently for their own nutrition (autotrophic) by the process of photosynthesis. In the green algae (*Chlorophyceae*) the cell walls are mainly of cellulose and are almost all found in freshwater or moist situations. Some are found living in symbiosis with fungi as lichens. Sexual reproduction is by the conjugation of two gametes, the male of which is always ciliated and the female, in some cases, is a non-motile egg-cell. *Pleurococcus*, a genus constituting a large part of the green slimy or powdery layer on many trees and wet wood out-of-doors, can be taken as an example.

Alnus. (*for*) A genus of north temperate hardwood trees. Two species of economic importance, *Alnus glutinosa*, known as Common Alder, and *Alnus incana*, known as Grey Alder. Very durable under fresh water. Formerly used for piles, sluice gates, etc. Easily pressure impregnated for other purposes. Has been used extensively for plywood.

Ambrosia beetles. (*ent*) Certain genera of the beetle subfamily SCOLYTINAE and the genus *Platypus* in the subfamily PLATYPODINAE possessing a special biology in which the adult beetle bores into the wood of a freshly-felled or damaged tree and takes in with it the spores or mycelium of a fungus which grow in the gallery. There is wide variation in the type of tunnel constructed and the behaviour of the larvae which hatch from eggs laid in the tunnel, but in general the larvae feed on the developing fungal mycelium. The larvae of some species remain in short side galleries, known as cradles, whilst others are more mobile. Ambrosia is an ancient word referring to the food of the gods, the beetle larvae first being described as feeding on 'a kind of ambrosia'. The fungi associated with each species of beetle are often specific and the relationship is one of symbiosis in that the development of the beetle and the fungus cannot take place without the other. Galleries made by ambrosia beetles may be identified by their straightness, lack of bore-dust and their dark-stained walls. They are sometimes referred to as pinholes or shotholes and the beetles as 'pinhole or shothole borers.' As the fungus requires moist conditions for growth it dies when the timber dries so that ambrosia beetles are never pests under such conditions. It is of interest that ambrosia beetles frequently attack wooden casks containing wine, spirits and fermenting fruit juices, obviously attracted to the casks by the odour produced. The chief genera are *Platypus*, *Anisandrus*, *Xyleborus*, *Gnathrotrychus*, *Pterocyclon*, *Trypodendron*, *Crossotarsus*, *Diapus* and *Xyloterinus*.

Amenability. (*proc*) To preservative treatment, is a measure of the ease with which a timber can be impregnated with preservatives. See *Permeable* and *Resistant*.

Annual ring. (*for*) A growth ring as shown in a transverse section of a trunk or branch corresponding to an annual increment of woody tissue.

Anobiidae. (*ent*) A family of Coleoptera, the larvae of the majority of whose 1,100 species are woodborers. The larvae are hook-shaped, larger at the front, and the terminal segments also enlarged. In the adult the head is inserted almost vertically downwards and the antennae have the three terminal segments enlarged. Two very important species attacking wood in Britain, and in other parts of the world, are, *Anobium punctatum*, the Common Furniture Beetle, and *Xestobium rufovillosum*, the Death Watch Beetle. *Ernobius mollis* is another anobiid found in buildings where the bark still adheres to softwood joists and rafters.

Anobium punctatum. (*ent*) A beetle in the family ANOBIIDAE popularly known as Common Furniture Beetle and whose larva is known, loosely, as 'woodworm'. One of the most important woodboring insects in the world, damaging a very wide range of timber species, and although formerly connected with old or antique woodwork, now known to infest fresh construction. About three-quarters of all the buildings in Britain whose timbers have been surveyed have been found to contain an infestation of this species, particularly in the softwood structural components of the roof void. In addition, joinery is often attacked. Eggs are laid in cracks on the wood surface and after fifteen to twenty-four days the resulting larva bores into the wood for three years or more. When the white, hook-shaped larva attains full development it bores towards the outer surface, and just beneath it constructs a pupal chamber. The pupal stage lasts from three to eight weeks when the adult beetle emerges from the wood, having bored a circular hole about one-sixteenth of an inch in diameter. *Anobium punctatum* is found throughout temperate Europe and temperate Asia, but also occurs in many other parts of the world where presumably it has been conveyed through commerce. Chief amongst these are New South Wales, Tasmania, New Zealand (where it is known as the house borer), South Africa and the Eastern States of the United States.

Antenna. (*ent*) pl. ae. One of a pair of mobile, segmented appendages articulating with the head in front of, or between, the eyes of an insect. They vary widely in form from being thread-like to comb, feather or book-like. They are usually organs of special sense but may be modified for other functions. Sometimes called horns or feelers.

Antiseptic. (*myc*) A substance which destroys harmful micro-organisms or renders them harmless.

Araucaria. (*for*) A genus of softwood trees from the southern hemisphere, generally but inappropriately called Pine suitably qualified. *Araucaria angustifolia* from Argentina and South Brazil is Parana Pine; *Araucaria araucana* from Chile is Chile Pine; *Araucaria bidwilli* from Australia is Bunya Pine, and *Araucaria cunninghamii* from Australia and New Guinea is Hoop Pine. Parana Pine is the most important softwood grown in South America. The sapwood is almost white but the heartwood is very variable, from pale to dark brown and sometimes partly bright red. The close textured and uniform wood is free from odour and contains little resin but is variable in respect of softness. Strength properties compare favourably with Baltic Redwood. It must be preserved if used out-of-doors. In South America it is used for almost every purpose for which a softwood is used in Europe. The veneered or turned knotty timber is sometimes seen as lampshades where the knots are translucent, glowing bright red.

Architrave. (*bldg*) A lining or covering to the frame within an opening such as a door or window in order to make a finished join with the wall surface.

Armillaria mellea. (*myc*) Honey Fungus. A basidiomycete fungus which attacks a great variety of plants ranging from forest trees to shrubs and herbaceous plants. The prominent rhizomorphs are bootlace-like.

Arris. (*bldg*) The sharp edge of any building element such as plaster or a brick. The upper edge of a building element.

Arris rail. (*bldg*) A fence rail of triangular cross-section fixed with an angle (or arris) uppermost.

Arthropoda. (*zoo*) The largest phylum of the animal kingdom whose members are characterised by possession of a segmented body with a chitinous external skeleton. Paired jointed appendages which are modified according to function are carried on a variable number of the segments. The more important classes included in this phylum are the Insecta (Insects), Crustacea (Lobsters, Shrimps, Crabs, Barnacles, etc.), Arachnida (Scorpions, Spiders, Mites, Ticks, etc.), Diplopoda (Millipedes), and Chilopoda (Centipedes).

Ascomycetes. (*myc*) A major group of the higher fungi (Eumycetes) characterised by the possession of a richly branched mycelium consisting of septate hyphae with chitinous cell walls. The asexual spores are developed in asci (singular is ascus) or tubular sporangia usually in groups of eight. Important examples are some common moulds, the mildews, as well as a number of wood-rotting species, e.g. *Chaetomium*.

A number of species of Ascomycetes are the cause of sapstain and soft rot.

Ascus. (*myc*) The sac-shaped end of a hypha on the hymenial layer of an ascomycete fungus which contains spores (usually eight).

Aseptic. (*myc*) Where all harmful micro-organisms have been killed or rendered harmless. Sterile.

Ash. (*for*) See *Fraxinus*.

Aspen. (*for*) See *Populus*.

Atlas cedar. (for) See *Cedrus*.

Atomizer. (*equip, myc*) A device for applying a spore or mycelial suspension as an inoculum in the form of fine droplets. For this purpose it must be capable of being detached from and sterilised separately from its rubber tubing and bulb, and a suitable sterile air filter must be interposed between the bulb and the atomiser.

Attractant. (*chem*) A substance or a physical condition which draws or attracts insects towards it from a distance. It is presumed that in many cases, eg attraction of ambrosia beetles to fermenting sap, the olfactory sense is exercised, but sight is obviously used when *Lyctus* beetles are attracted to light.

Autoclave. (*equip, myc*) A thick-walled vessel with tightly fitting and safety-locking door in which temperature and pressure can be controlled in order to sterilise medium and apparatus.

Autoclave. (*equip*) See *Pressure chamber.*

Auto immersion unit. (*equip*) A tank equipped to automatically and totally immerse timber in preservative solution. The speed can be regulated so as to give the desired immersion period.

B

Bacillus. (*bac*) Genus of rod-shaped, spore-forming bacteria which includes a number of physiologically important species. *Bacillus amylobacter* decomposes cellulose. *B. butyricus* forms butyric acid and *Bacillus saccharobutyricus* fixes atmospheric nitrogen.

Bacteria. (*bac*) A group of the Thallophyta the members of which are of extraordinarily small size, the smallest having a diameter of less than one thousandth of a millimetre. They do not possess chlorophyll and the majority are unicellular. In shape they are spherical, rod-shaped or having the form of a spiral, but a number have branched cells and some are united to form filaments. A central nucleus is absent but the presence of nuclear material can be demonstrated as scattered granules. Many bacteria possess delicate protoplasmic cilia at certain stages of development which allows movement. The importance of the presence of bacteria in the early stages of decay has recently been demonstrated.

Baluster. (*bldg*) A single vertical component of a balustrade.

Balustrade. (*bldg*) The infilling from handrail to floor level to the sides of a flight of stairs.

Bandage. (*proc*) See *Ground line treatment.*

Bankia. (*zoo*) A genus of Mollusca related to *Teredo*, the species of which tunnel into wood submerged or floating in the sea. See *Marine borers.*

Bark. (*bot*) Protective layer surrounding plant stems. A composite tissue but largely consisting of cork. See *Cork.*

Bark beetles. (*ent*) Beetles in the family SCOLYTIDAE, closely allied to the weevils (CURCULIONIDAE). Sometimes Ambrosia beetles are included. They range from one-sixteenth to just over quarter of an inch in length, and the legless larvae inhabit tunnels which they themselves have made. Bark beetles lay their eggs in the bark or in the cambium layer of the host tree, in a tunnel excavated for the purpose. This is known as the mother-tunnel and may be extremely short. The eggs are usually laid in niches cut in the wall. Pairing tunnels are sometimes bored, as are ventilating tunnels also. Occasionally, the mother-tunnel is two-armed as in *Hylesinus* and *Hylastinus.* The larval tunnels usually radiate outwards from the egg tunnel, increasing in width as they extend and ending with the pupal chamber. Sometimes the latter curve inwards to the sapwood.

The general pattern of the brood gallery (the whole complex of galleries) is classified into various groups and may frequently be used for identification of the species concerned.

Barking. (*proc*) The act of removal of bark from a log or tree.

Basidiocarp. (*myc*) Sporophore. The structure bearing the hymenial layer from which the basidia arise in the BASIDIOMYCETES.

Basidiomycetes. (*myc*) A group of the higher fungi in which the spores are produced on club-shaped structures known as basidia. Usually four spores are borne on each basidium, each on a slender stalk from which they are violently discharged when ripe. The basidia are formed in a compact superficial layer known as the hymenium and the sub-division of the group into families is made on the shape and position which it takes. Most of the larger and more important wood-destroying fungi belong to the BASIDIOMYCETES and to four of its principal families as follows:

1. Hymenium exposed freely on a flat skin-like surface. THELEPHORACEAE.
2. Hymenium on surface of spine-like outgrowths. HYDNACEAE.
3. Hymenium lining inside of pores or tubes. POLYPORACEAE.
4. Hymenium on plate-like gills underneath a cap-shaped pileus of mushroom type. AGARICACEAE.

Basidium. (*myc*) A special cell of the hymenial layer of a basidiomycete fungus which bears spores, usually four, on stalk-like sterigmata.

Basswood. (*for*) See *Tilia.*

Bast. (*bot*) The inner bark or phloem tissue.

Beech. (*for*) See *Fagus.*

Bethel treatment. (*proc*) See *Full-cell process.* Patented by John Bethel in 1838.

Betula. (*for*) A genus of north temperate hardwood trees known as birch. Two common European species are *Betula pubescens,* White, or Black Birch, and *Betula verrucosa* or *B. pendula,* Silver Birch or Common Birch. The timber is almost white in colour. It is rapidly attacked by fungi out-of-doors, but is easily pressure impregnated. It is susceptible to attack by *Anobium punctatum* indoors. *Betula lutae,* Canadian Yellow Birch has been widely used in joinery and stained red has marked similarity to Mahogany.

Bio-assay. (*bio*) The estimation of efficiency of a process or a material for a particular purpose by the use of living organisms.

Biocide. (*chem*) A substance which kills living organisms.

Biodeterioration. (*bio*) Biodecay. The loss of one or more desirable properties possessed by a substance or manufactured article through the action of living organisms. Biodeterioration affects a wide range of materials, goods and structures. Some examples are the attack on wood, paper, textiles, and grain by fungi and insects, the blocking and corrosion of pipelines by bacteria, the growth of fungi and algae on concrete and painted surfaces, the fouling of ships and marine structures by barnacles and molluscs, the destruction of stored products by rodents and the hazards caused to aircraft by birds.

Biodeterioration Information Centre. (*name*) Set up at the University of

Aston in Birmingham 1965 aided by grant from Office for Scientific and Technical Information (UK Department of Education and Science). It collects all literature on deterioration of materials of economic importance by living organisms and in this work receives much help from a large number of biodeterioration experts ('co-operating specialists') who send in references from many countries. Each item of literature is indexed with appropriate keywords and recorded in a punch card index system. Retrieval from this is the basis for the rapid answer of enquiries. A classified bibliography of the literature held at the Centre (International Biodeterioration Bulletin Reference Index Supplement) (IBBRIS), is published four times a year.

Biodeteriorgen. (*bio*) A living organism causing the loss of one or more desirable properties possessed by a substance or manufactured article.

Biological attack. (*bio*) The damage done (to wood) by living organisms which either use wood or wood constituents as a nutrient, or for shelter, and by so doing destroy it either wholly or in part. The principal organisms taking part in the biological attack of timber are Fungi, Bacteria, Insecta, Crustacea and Mollusca.

Birch, Black, Canadian Yellow, Common, Silver, White. (*for*) See *Betula.*

Blackheart. (*for*) An abnormal black or dark brown discoloration that may occur in the heartwood of certain timber species. Not necessarily associated with fungal decay.

Bleeding. (*chem*) The exudation of liquid preservative from treated wood. The exudate may evaporate, remain liquid, or harden into a semi-solid or solid material.

Blooming. (*chem*) The formation of crystals on the surface of wood treated with preservative; caused by migration to the surface of one or more of the components and the subsequent evaporation of the solvent.

Blue stain. (*myc*) A common form of Sapstain of bluish colour usually confined to the sapwood of coniferous timber. It is brought about by the action of fungi, commonly occurring genera of which are *Ceratocystis* (*Ceratostomella*) and *Lasiodiplodia*, at a timber moisture content above 20 to 25%, maintained for at least ten days, and at a temperature suitable for fungal growth. See *Sapstain.*

Bog oak. (*bot*) Oak which has become stained black due to having been buried in bogs for some hundreds of years. Commonly used for small fancy articles especially in Ireland.

Boiling under vacuum. (*proc*) See *Boulton process.*

Boiling without vacuum. (*proc*) A treatment for unseasoned timber in order to make it suitable for the pressure impregnation of preservative. It consists of boiling in oil preservative at atmospheric pressure and above 100°C.

Bole. (*for*) The main stem or trunk of a tree usually of minimum diameter of eight inches.

Boliden K33. (*chem*) A proprietary brand of copper/chrome/arsenic wood preservative.

Bolt-hole-treater. (*equip*) An implement for applying a preservative under pressure to holes bored into timber.

Borax. (*chem*) $Na_2B_4O_7.\ 10H_2O$ (381·4). White crystalline salt, about 2% soluble in water at 20°C. See *Boric acid.*

Bore hole. (*bot, zoo*) (a) A hole in a cell wall caused by the enzyme action of an advancing hypha of a wood-inhabiting fungus. (b) A hole or tunnel in woody tissue caused by a wood-boring animal such as the larva of a beetle.

Boric acid. (*chem*) H_3BO_3 (61·8) A white crystalline solid, very slightly soluble in cold water. The borate ion has useful fungicidal, insecticidal and fire retardant properties so that boric acid and its sodium salts are included in many formulations applied to wood.

One of the most important types of preservative for diffusion treatment consists of a concentrated solution of sodium octaborate, $Na_2B_8O_{13}.\ 4H_2O$ in water. This is usually heated to between 40°C and 60°C to maintain the correct concentration of borate ion in solution. The equivalent solution can be prepared from a mixture of borax and boric acid in the weight ratio 1·54:1.

Bostrychidae. (*ent*) A family of beetles which, together with the closely allied LYCTIDAE, are known as powder post beetles. About 550 species are known from the whole world and in all cases the larvae tunnel into wood and often cause great destruction, especially in tropical and subtropical areas. The body is often truncated posteriorly and armed with spines. The larvae are crescentic and resemble those of the ANOBIIDAE, except that the head is much reduced in size.

Boucherie process. (*proc*) An early form of sap displacement process patented in 1838 in France. In its later forms, a water-based preservative, usually copper sulphate, was fed to the butt-end of horizontal green poles from a tank 25–30 feet above the poles. Various ways of sealing the connecting hose to the pole were used. The hydrostatic pressure forced preservative along the length of the pole, displacing the natural sap. In a few days, complete impregnation of the sapwood was achieved.

Boulton process. (*proc*) A treatment patented in 1879 for unseasoned and partially seasoned wood in order to make it suitable for the pressure impregnation of preservative. It consists of heating the timber in oil, under vacuum, in order to expel the water.

Bound moisture. (*bot*) Moisture intimately associated with the cell wall which is not freely given up.

Bréant. (*proc*) The first patentee, in 1831, of the use of hydraulic pressure for impregnation of wood by preservative solution, and the patentee of the use of vacuum in 1837.

Bright. (*for*) Timber free from discoloration or from any dullness due to the presence of wood destroying fungi.

British Wood Preserving Association or BWPA. (*name*) Was founded in

1930 for the purpose of collecting, promoting and spreading knowledge of all methods of wood preservation, the protection of wood against fire and the promotion of standard preservative specifications.

It sponsors scientific research through its members into the use of preservatives and fire retardants and makes the results available to all enquirers by the publication of leaflets, a free advisory technical service and specialist lectures. The advice it gives is impartial. One important aim is making known the advantages derived from the use of preserved timber in the interests of both the consumer and the national economy.

Membership consists of learned societies and research bodies at home and overseas, manufacturers of preservatives and fire retardants, firms operating various types of treating plants, specialists in the remedial and curative treatment of timber, plant manufacturers, architects, builders, consultants and timber users. Its committees deal with service records and service tests, specifications, technical problems, and organic solvent preservatives. Liaison with government departments, the principal consuming industries, Forest Products Research Laboratory, British Standards Institution and Timber Research & Development Association, is maintained. An active part is played in the international field through such organisations as OECD, IUFRO, CEN, and FAO.

An Annual Convention Record is published consisting of about seven presented papers, together with the discussion. A monthly news sheet is published as also are standards for wood preservatives. Every two years a comprehensive Wood Preservatives and Fire Retardants Register and an annual Preservation Supplement are issued. Special reports are also prepared in collaboration with such organisations as TRADA and FPRL.

Brittleheart. (*for*) Wood in which abnormal brittleness occurs due to natural compression failures in the fibres. It is of common occurrence in certain tropical light hardwoods.

Broad-leaved. (*for*) Trees having broad leaves as opposed to needle or narrow leaves. Generally they are thought of as the deciduous angiosperms but broad-leaved evergreens are also common.

Brown oak. (*myc*) See *Fistulina hepatica.*

Brown rot. (*myc*) A decay of wood caused by a fungus which has utilised the cellulose and the associated pentosans whilst the lignin is left in a more or less unchanged condition. The wood turns brown in colour. The term 'destruction rot' has also been proposed. In the case of a brown-rot causing fungus, the cell wall of the wood does not appear to be thinned appreciably until a very late stage in decay. Even when practically all the cellulose lattice has been removed, the residual lignin preserves the form of the original cell wall.

Examples of fungi causing brown-rot are *Trametes serialis*, *Coniophora cerebella*, *Lenzites trabea*, *Merulius lacrymans* and *Poria xantha.*

Brush treatment. (*proc*) The application of a preservative to timber by means of a brush.

Bullnose. (*bldg*) Generally, any rounded edge or end such as of a brick, step or sill. The rounding of an arris.

Bunya pine. (*for*) See *Araucaria*.

Buprestidae. (*ent*) A family of Coleoptera whose 15,000 or so species are essentially of tropical distribution although twelve species occur in Britain. They are amongst the most brilliantly coloured of all insects, metallic greens, blues, and reds being usual. They are often used as jewels. The larvae, the majority of which bore flattened galleries in the outer sapwood of trees, are easily distinguished by their very large flattened prothorax into which the head is withdrawn, vestigial or absent legs, and slender hind body. *Buprestis aurulenta* is sometimes found in Britain emerging from Douglas Fir originating from North America. See plate 13.

Burnett process. (*proc*) A process patented in 1838 by Sir William Burnett using 1·5 per cent zinc chloride solution and a pressure of 125 to 150 lbf/in^2. The air having been previously exhausted from the retort. Of historical interest only.

Butt. (*for*) The base of the main trunk of a tree.

Butt treatment. (*proc*) A preservative treatment applied only to the butt end of posts or poles, usually by the hot and cold open tank process.

C

Cambium. (*bot*) The tissue producing secondary growth in plants. An actively dividing tissue in a tree occurring as a delicate sheath between the wood and the bark throughout the bole and the branches, producing cells to the inside which form woody tissue (Xylem) and to the outside which form bark tissue (Phloem).

Camponotus herculaneus. (*ent*) Carpenter Ant. A species of ant (FORMICIDAE) which injures outdoor woodwork, including building timbers and sometimes living trees. Large irregular cavities are gnawed out, the softer spring wood being extracted leaving the harder summer wood. Infested wood is characterised by the absence of bore dust, faecal pellets and fungal growth. Piles of fibrous borings removed to the outside of the wood are an indication of *Camponotus* attack. It does not occur in Britain but is sometimes introduced accidentally. It is often of economic importance in North America.

Canker. (*for*) A malformation of the bark of a living tree due to the death of the underlying tissue. A number of species of wood-destroying fungi are known to be the cause of various sorts of canker but, in addition, there are many other reasons.

Capillarity. (*phy*) Capillary action. A phenomenon observed when a tube of fine bore is inserted into the surface of a liquid when, due to forces produced by molecular attraction, the surface of the liquid is seen either to rise or fall in the tube. It is the main cause of end-grain absorption in timber when immersed in a preservative at atmospheric pressure.

Carbohydrate. (*chem*) A group of chemical compounds composed of carbon, hydrogen and oxygen and generally described by the formula $C_x(H_2O)_y$. The group contains sugars which, in the leaves of trees, are formed by photosynthesis from carbon dioxide and water; and polysaccharides such as cellulose, hemicelluloses and starch whose molecules consist of a number of sugar units linked together.

Card process. (*proc*) A full-cell preservative process using 80 per cent zinc chloride solution and 20 per cent coal-tar creosote. Not now generally used.

Carpenter ant. (*ent*) See *Camponotus herculaneus*.

Carpenter bees. (*ent*) See XYLOCOPIDAE. In Britain a number of the so-called Carpenter Bees classified in the APIDAE occur, but none bore into sound wood, although the larval cells are often found in short rows in tunnels excavated in fungally decayed wood. The rotten wood merely acts as a receptacle for the cells. In *Megachile* these are lined with pieces cut from leaves. Another common species is *Osmia rufa* but this species will place its larval cells in anything convenient such as key-holes, tyre-treads, loose sand, etc.

Casein. (*chem*) The main protein obtained from milk by souring. It was used extensively as an ingredient of adhesives for wood and wood-based board materials but has now been largely superseded by synthetic resin adhesives. Casein-glued plywood can be very susceptibe to insect attack.

Casement window. (*bldg*) A window in which one or more lights are hinged to open (BS565).

Cedar. (*for*) See *Cedrus.*

Cedar of Lebanon. (*for*) See *Cedrus.*

Cedrela. (*for*) A genus of timber-producing trees in the family MELIACEAE usually with fragrant odour and mostly from Central and South America, but also from India.

Cedrus. (*for*) A genus of softwood trees from the north temperate Old World, known as cedar. *Cedrus atlantica* is known as Atlas Cedar; *Cedrus deodara* as Deodar is an important timber species in northern India; *Cedrus libani* is well known as Cedar of Lebanon. Timber of these species is usually only available in Britain from trees planted for ornamental purposes. The fragrance of the wood, however, has caused a large number of timbers to be called cedar suitably qualified. See *Chamaecyparis, Cryptomeria, Juniperus, Libocedrus, Thuja and Cedrela.*

Celcure. (*chem*) A proprietary brand of copper/chrome waterborne wood preservative. See BS.3452:1962.

Celcure AP. (*chem*) A proprietary brand of copper/chrome/arsenic wood preservative. See BS.4072:1966.

Cell. (*bio*) Microscopic component of which plant and animal tissue is built up. The living cell contains a complex of organelles dispersed in protoplasm.

Cell wall. (*bio*) The rigid surrounding of plant cells, composed of one to three layers. No such rigid wall is present in animal cells.

Cellar fungus. (*myc*) See *Coniophora cerebella.*

Cellulase. (*bio, chem*) The enzyme responsible for the degradation of cellulose and therefore of immense importance in the carbon cycle. It catalyses the hydrolysis of cellulose. It has been stated that cellulase is a major limitation to the usefulness of wood, paper, pulp, cotton, rayon, cellophane, and a host of other cellulosic materials of great and diverse utility.

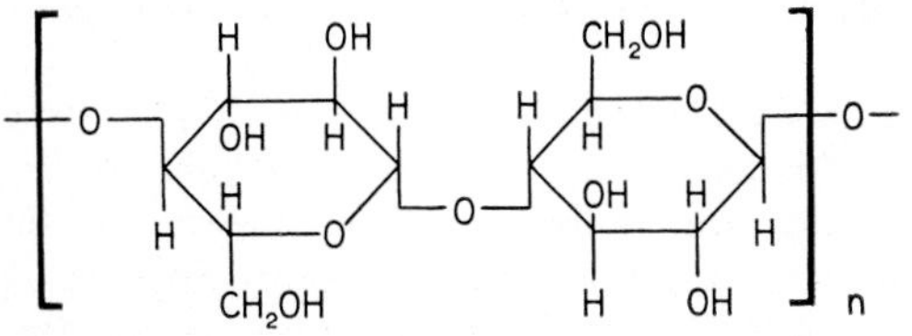

Cellulose. (*bio, chem*) A carbohydrate substance or group of closely-related substances forming the main chemical constituent of the cell walls in wood. It is based on polymerised glucose units $(C_6H_{10}O_5)_n$, where n may be up to about 3,000. The molecular chains are unbranched, but appear to be bound laterally into fibrils. If wood cellulose is treated with nitric or acetic acid, it produces a soluble compound utilised by many industries as a basic material.

Cerambycidae. (*ent*) A family of insects in the order Coleoptera (beetles) and in the sub-order Polyphaga, sometimes referred to as LONGICORNIA. About 20,000 named species are known from the whole world and are found wherever trees or bushes grow, and wherever timber is transported or used. They are characterised by the long antennae which are usually at least two-thirds as long as the whole body and are inserted on prominent tubercles which allows them to be flexed backwards over the body. The larval stage is most often passed in dead or decaying trees but a few species are to be found in the sap or heartwood of living trees. Many species inhabit buildings and other timber, and are of economic importance. In Europe the species *Hylotrupes bajulus* is outstanding in this respect.

Chaetomium globosum. (*myc*) A well-known species of the cellulose-attacking ascomycetes and one of the species of microfungi causing the decay of non-durable hardwoods known as Soft Rot. See Plates 3 and 5.

Chamaecyparis. (*for*) A genus of softwood trees from North America and Japan. *Chamaecyparis lawsonia* from Oregon and California is Port Orford Cedar. This species was formerly placed in the genus *Cupressus* and was also known as Lawson's Cypress. *Chamaecyparis nootkatensis* from the Pacific Coast region of North America is Yellow Cedar and *Chamaecyparis thyoides* from USA is Southern White Cedar. See notes under *Cedrus*. The fairly hard and close-grained timber possesses a strong fragrant scent. It lays some claim to durability as it has been used in North America for railway sleepers and shingles.

Charge. (*proc*) A unit of timber for batch treatment. The quantity of timber which can be dip or pressure treated by a particular process at any one time, this obviously depending upon the dimensions of the treatment cylinder or tank.

Check. (*bldg*) See *Rabbet*.

Check. (*for*) A split in the surface of a piece of timber, usually resulting from shrinkage during seasoning. End-checking is often prevented by applying a water impermeable barrier paint or paste to the ends of round or sawn items.

Chelura. (*zoo*) A genus of AMPHIPODA in the Crustacea associated with gribble damaged timber submerged or floating in the sea. See *Limnoria*.

Chemical attack. (*chem*) The damage done (to wood) as a consequence of the action of chemicals such as acids and alkalies or oxidising agents. Generally speaking most timbers are fairly resistant in this regard.

Chestnut, Sweet. (*for*) *Castanea sativa*. Sweet or Spanish Chestnut. Used extensively for fencing, for which purpose the tree is often coppiced.

Chile Pine. (*for*) See *Araucaria.*

Chlordane. (*chem*) $C_{10}H_6Cl_8$ (410) The common name for 1,2,4,5,6,7, 10,10-octa-chloro-4,7,8,9-tetrahydro-4,7-methyleneindane. Its insecticidal properties were first described by Kearns, C. W. *et al.*, *J. econ. Ent.*, 1945, **38**, 661; an independent discovery was reported by Riemschneider, R., *Chim. et Ind.*, 1950, **64**, 695, under the code number 'M 140'. It was introduced, in 1945, by the Velsicol Corporation under the code number 'Velsicol 1068' and the trade name 'Octachlor', protected by BP618,432.

For its manufacture, hexachlorocyclopentadiene, made by the chlorination of pentanes: USP2,509,160 or by the action of sodium hydrochlorite on cyclopentadiene: USP2,606,910, is condensed with cyclopentadiene to produce chlordene, $C_{10}H_6Cl_6$; chlordene is further chlorinated to chlordane.

The technical product is a viscous amber-coloured liquid of d^{25} 1·59 to 1·63, n_D^{25} 0·56, viscosity 75 to 120 centistokes at 130°F. It is insoluble in water but soluble in most organic solvents including petroleum oils. The refined product has a vp of 1×10^{-5} mm Hg at 25°C.

Cl
Cl Cl
CCl₂
Cl Cl
Cl

Technical chlordane consists of 60 to 75 per cent isomers of chlordane and 25 to 40 per cent of related compounds including two isomers of heptachlor, *q.v.*, and one each of enneachloro- and decachloro-dicyclopentadiene: Riemschneider, R., *Wld. Rev. Pest Control*, 1963, **2**(4), 29. All have the endo configuration. Two isomers of octachlorodi-cyclopentadiene have been isolated from chlordane, both of mp 103 to 105°C, of which alpha-chlordane is the endo-cis and beta-chlordane is the endo-trans isomer. The alpha isomer is much more readily dehydrochlorinated than the beta isomer. The commercial product known as gamma-chlordane is substantially the alpha isomer, though this name has recently been applied to the 2,2,4,5,6,7,8,8-octachloro isomer, mp 131°C, for which greater insecticidal and lower mammalian activity is claimed.

Chlordane is a persistent, non-systemic stomach and contact insecticide: non-phytocidal at insecticidal concentrations. It is of especial interest in termite control by soil treatment.

The acute oral LD_{50} of chlordane for rats is 457 to 590 mg/kg; a high vapour toxicity to mice reported by Frings, H. and O'Tousa, J. E., is due to unreacted hexachlorocyclopentadiene now reduced to insignificant amounts in the technical product. Rats fed for 104 weeks on a diet containing 150 ppm gamma-chlordane suffered no higher mortality than the controls, but histo-pathological changes in the liver were apparent.

Product analysis is by the determination of total chlorine by the Stepanow method: AOAC Methods. Residues may be determined (a) by reaction with diethanolamine and methanolic KOH, the purple colour being measured at 521 mμ: or (b) by gas chromatography.

Chloropicrin. (*chem*) $CCl_3.NO_2$ (164·5) The common name for trichloronitromethane, also known as nitrochloroform. Its use as an insecticide was protected in 1908 by BP2387.

Made by the action of hypochlorites and steam on calcium picrate, it is a colourless liquid of mp—64°C, bp 112·4°C, vp 5·7 mm Hg at 0°C, 23·8 mm Hg at 25°C; d_4^{20} 1·656; n_D^{20} 1·595. Its solubility in water is 2·27 g/litre at 0°C. and it is miscible with acetone, benzene, carbon tetrachloride, ether, methanol.

Chloropicrin is non-inflammable and chemically rather inert. It is non-corrosive to copper, brass and bronze but attacks iron, zinc and other light metals; the formation of a protective coating permits storage in iron or galvanised containers.

It is an insecticide used for the fumigation of storage grain and timber buildings against termites.

It is used at 2 to 5-lb/1,000 ft^3; and is more generally used in fumigants as a warning gas on account of the intense irritation of the mucous membranes, serving to reduce hazards.

For identification, add one drop to 5 ml 20 per cent sodium sulphide, close tube, the odour of chloropicrin disappears. For detection: (a) pass air through sodium ethylate solution and test for nitrites; (b) pass air through 4 per cent sodium nitrate in ethanol, test for chlorides.

For determination in air: collect in isopropanol, oxidise and neutralise with HCl, react the liberated nitrite with sulphanilic acid, couple with N-1-naphthylethylenediamine and measure at 540 mμ.

Chromate. (*chem*) CrO_4^{2-} The simplest oxy-anion of hexavalent chromium. Salts of both this ion and of the dimerised form dichromate, $Cr_2O_7^{2-}$ are used in many water-borne wood preservatives because of their ability to form insoluble (fixed) compounds in wood with various other ions, principally copper, zinc, arsenate and phosphate. Chromates and dichromates are usually yellow or orange-red, water soluble and crystalline.

In preservative solutions they also act as corrosion inhibitors and therefore prevent rapid wastage of the metals used in treatment plant construction.

Chronic dose. (*tox*) An amount of a substance which when administered causes a long-lasting harmful effect.

Cladosporium cladosporioides. (*myc*) A species of fungus producing a mould, the spores of which are used in the preparation of a mixed spore suspension in the test for mould and mildew resistance of manufactured building materials. See BS1982: 1968.

Clamp-connection. (*myc*) A swelling on the side of the hypha of certain basidiomycete fungi through which cell material is passed for fusion with cell material of the adjacent cell.

Clean. (*for*) (timber) Knots absent.

Clear. (*for*) Timber free from all visible defects and imperfections.

Cleridae. (*ent*) A family of Coleoptera of which the nearly 3,000 species are mainly tropical and are of brilliant coloration. The larvae of many species are predaceous on wood-boring larvae. The larva of *Korynetes caeruleus* preys on the larvae of *Anobium punctatum* and *Xestobium rufovillosum*.

Close-couple or Couple-close. (*bldg*) A descriptive term of a roof-structure in which the common rafters of opposite sides are joined by a tie-beam at wall-plate level.

Close piling. (*proc*) A method of stacking sawn timber without intervening air spaces. Usually recommended when dip-diffusion preservatives have been used. Also known as close stacking.

Cobra process. (*proc*) An *in-situ* process for preservative treatment of transmission poles where the soil is removed from the base of the pole and a portable apparatus fitted. On striking a knob, 2 to 3 grams of preservative salt are forced into the wood through a hollow needle to a depth of about 3 cm. Patented 1922 by Laube (British) and Schmittutz 1924 (German).

Coleoptera. (*ent*) Beetles. A very highly developed order of insects which contains about 275,000 different species, thus, also being the largest order in the animal kingdom. They are characterised by the forewings being modified into horny or leathery elytra which almost always meet to form a straight mid-dorsal suture whereas the hindwings are membraneous and fold beneath them. The hindwings are sometimes absent. The mouthparts are adapted for biting, the large prothorax is mobile, and a complete metamorphosis takes place. Some of the most minute, and some of the largest, of all living insects belong to this order. The biology of beetles is most diverse and no other order of insects have invaded land, water, and air to anything like the same extent. Almost every form of organic matter (except living animals) is utilised as food by beetles. Many families possess wood-boring habits, the most important of which are BUPRESTIDAE, ANOBIIDAE, BOSTRICHIDAE, LYCTIDAE, LYMEXYLIDAE, OEDEMERIDAE, CERAMBYCIDAE, CURCULIONIDAE, SCOLYTIDAE and PLATYPODIDAE. A number of species are amongst the most important of wood-destroying animals.

Collapse. (*phy*) A defect of wood which may occur either during too rapid kiln-drying or during pressure treatment with preservatives. In either case, the stresses within the timber cause severe distortion of cells and a change in overall shape and size of the item. The tendency for collapse to occur usually determines the maximum pressure that can be applied during preservation of a timber species. If it does occur, the penetration of preservative is adversely affected, in addition to the distortion of the timber.

Collar. (*bldg*) An increase in the outside diameter or a decrease in the inside diameter of a pipe so that two adjacent pipes can bear upon or fit into each other.

Collar beam. (*bldg*) Also known as a top beam or a span piece. A horizontal tie-beam in a roof. See *Collar Beam Roof*.

Collar beam roof. (*bldg*) The structural timbering of a roof in which the common rafters are joined half-way up their length by horizontal tie-beams. More headroom is given by this type of roof than by a close-couple roof.

Common furniture beetle. (*ent*) See *Anobium punctatum*.

Common rafter. (*bldg*) The sloping rafter carrying the roof covering. In a single roof it is fixed at its base to the wall plate and to the ridge at the top. When principal rafters, connected by purlins, are present it lies on top of the purlins.

Compression wood. (*bot*) Modified woody tissue found in the timber converted from coniferous trees which have grown in a leaning position often on steep slopes. It occurs on the lower, downhill, side of the tree and is characterised by its dark colour and lack of contrast between early and late wood. Although exceptionally dense the strength properties are poor. The most important character, however, is the abnormal longitudinal shrinkage on drying, which is commonly up to ten times the usual figure and may cause excessive bowing.

Conditioning. (*proc*) A preliminary treatment given to unseasoned or imperfectly seasoned wood to remove water in order to improve its properties in respect of the penetration and absorption of wood preservatives.

Conidiophore. (*myc*) The modified hypha of a fungus bearing the conidia.

Conidium. (*myc*) pl. Conidia. A thin-walled fungal spore born terminally on a specialised hypha or conidiophore. Produced by constriction of part of a hypha *e.g.* spores of a green mould.

Coniferous. (*for*) Applied to cone bearing trees or timber converted from them.

Coniophora cerebella. (*myc*) Cellar Fungus. A very common wood-rotting fungus in buildings, as well as out-of-doors, causing a decay usually referred to as 'Wet Rot'. Hard and softwoods are attacked and the decay is characterised by the wood breaking up into small pieces (much smaller than those in the case of attack by *Merulius lacrymans*), and the longitudinal cracks, along the grain are much larger than those at right-angles. The decayed wood is brown or almost black and in many cases the decay in a joist is confined to the inside whilst a thin layer of apparently sound wood surrounds it. Concavities in the wood, however, point to its presence. The mycelium, whose growth is confined to damp areas, and usually sparse, forms, at most, a thin yellowish-brown skin. The thin strands become dark in colour and never penetrate brickwork although they sometimes produce fern-like patterns on damp plaster. The inconspicuous sporophore, olive-green when fresh and soon darkening, is skin-like with small knobs and the spores are pale olive-brown. See Plates 5 and 14.

Contact angle. (*phy*) The angle made by the surface of a drop of liquid resting on a solid substrate. Of use in assessing the initial efficacy of water repellent solutions for wood. The highest water repellent properties are possessed by a liquid showing a constant contact angle higher than 90°.

Contamination. (*bio*) The accidental introduction of a harmful micro-organism into a culture. The accidental introduction of a harmful chemical substance into an environment.

Conversion. (*for*) The process of sawing a log up into timber.

Copper-chrome-arsenate. (*chem*) CCA. A class of water-borne preservatives applied by a pressure process and in wide use throughout the world. Patented by S. Kamesam in 1933 and known as 'Ascu', the first CCA consisted of potassium dichromate ($K_2Cr_2O_7$), 5 parts, copper sulphate ($CuSO_4.5H_2O$), 3 parts, and arsenic pentoxide ($As_2O_5.2H_2O$), 1 part. This formulation resembled the acid copper chromate known as 'Celcure' which had been in use since 1927, but had some of the copper sulphate replaced by arsenic pentoxide. 'Ascu' has been in wide use in India since 1938. In 1942, Bell Telephone Company published a survey of pole treatment using a preservative called 'Greensalt' which was similar to Ascu. From 1936 to 1948 the Boliden Mining Company in Sweden produced a zinc-chrome-arsenate preservative called 'Boliden' BIS and S25 which was a metal-chrome-arsenate containing both zinc and copper. BIS and S25 contained rather more of the toxic zinc, copper and arsenic, but less chromium, and for this reason were more easily leached than was Ascu. Boliden then followed with a copper-chrome-arsenic in the form of oxides. In the meantime, in the United Kingdom Celcure 'A' and Tanalith 'C', two new copper-chrome-arsenates became available.

A feature of CCA preservatives is the high degree of fixation of the components which occurs in wood soon after treatment. This is mainly due to their chromium content. The biocidal effect is conferred by the copper and arsenic.

Copper naphthenate. (*chem*) The copper derivatives of naphthenic acids (naturally occurring carboxylic acids extracted from petroleum oils). Manufactured by the direct interaction between soluble copper salts with naphthenic acid. According to British Standard 3769:1964, the material shall contain metallic naphthenate equivalent to not less than 10 per cent of metallic copper when determined by a prescribed method. Copper naphthenate is a solid or very viscous liquid of deep green colour but which is liquid when heated to 100°C. Usually possesses an unpleasant odour; vp less than 0·001 mm Hg at 100°C. Practically insoluble in water, moderately soluble in petroleum oils and soluble in most organic solvents.

Solutions of copper napthenate have been in long use as wood preservatives and have been made well-known under the trademark 'Cuprinol'.

Copper naphthenate-based solution. (*chem*) A solution of copper naphthenate (see British Standard 3769:1964) in a petroleum or coal tar distillate with a boiling range between 140°C (284°F) and 270°C (518°F) when tested by the method given in BS3769 (appendix j). The metallic copper content shall be not less than 2·75 per cent by weight. According to BWPA Standard 106, the flash-point of the solution shall not be lower than 33·9°C (93°F). This standard should be consulted for further details, methods of analysis and estimation of preservative in treated wood.

This bright green solution has enjoyed a wide application as a wood preservative applied by non-pressure methods.

Copper naphthenate concentrates. (*chem*) Two concentrates of copper naphthenate are recognised by British Standard 3769:1964, 8 per cent and 6 per cent copper content respectively. The solvent used is a petroleum distillate. The 8 per cent copper concentrate is a semi-solid or a liquid, but is liquid when heated to 100°C. The 6 per cent copper concentrate is liquid.

Copper sulphate. (*chem*) $CuSO_4$. $5H_2O$ (249·5) The common name for the pentahydrate of cupric sulphate, also known as bluestone, blue vitriol, blue copperas. The fungicidal activity of soluble copper salts was discovered, in 1807, by Prévost.

Copper sulphate is prepared by the action of sulphuric acid on copper or copper oxide and forms blue crystals of d^{15} 2·286. Its solubility in water is 31·6 g/100 ml at 0°C and it is insoluble in ethanol. The crystals are slightly efflorescent in air and lose water of crystallisation at 110°C to form the white monohydrate, $CuSO_4.H_2O$. Copper sulphate solutions are strongly corrosive to iron and galvanised iron.

Copper sulphate is an important constituent of the water-borne copper-chrome-arsenate compositions. To British Standard 4072:1966, copper (calculated $CuSO_4.5H_2O$) is present to the extent of 32·6 per cent w/w as nominal and 30 per cent w/w as minimum in type 1 composition and 35 per cent w/w and 31·5 per cent w/w respectively in type 2.

For method of determination of copper content of formulated copper/chrome/arsenic mixture (and for determination of amount of copper retained in treated timber) see above BS.

Coppice. (*for*) Wood of small growth for periodical cutting, sprouting from cut stumps.

Coptotermes. (*ent*) A genus of termites in the RHINOTERMITIDAE (dampwood and subterranean) including the following major pests of buildings, *C. acinaciformis* in Australia and New Zealand, *C. amani* in East Africa, *C. ceylonicus* in Ceylon, *C. crassus* in Central America, *C. exiguus* in Ceylon, *C. formosanus* in the Far East, Hawaii, Guam, Midway and Marshall Islands, Ceylon, South Africa and Gulf coast of USA, *C. grandiceps* in Solomon Islands, *C. havilandi* in Malayan Region, Madagascar, Mauritius and Caribbean, *C. niger* in Central America, *C. pamuae* in Solomon Islands, *C. sjostedti* in West Africa,

C. testaceus in Caribbean and South America, *C. travians* in the Malayan region, *C. truncatus* in Madagascar and *C. vastator* in Philippines.

Cork. (*bot*) Protective tissue, impermeable to liquids and gases, surrounding plant stems. A constituent of bark. See *Bark*.

Corrosion. (*chem*) The gradual destructive action on a product or substance by a chemical agent. Preserved wood likely to come into contact with certain metals, e.g. aluminium, must not exhibit this.

Corrosive sublimate. (*chem*) See *Mercuric chloride*.

Cossidae. (*ent*) A family of moths containing several species injurious to timber in Europe, including the Goat Moth, *Cossus cossus*, and the Wood Leopard, *Zeuzera pyrina*.

Cottonwood. (*for*) See *Populus*.

Creep. (*phy*) The slow non-elastic deformation of wood which usually follows rapid elastic deformation when the wood is placed under stress. On removal of the stress, elastic recovery is again rapid but recovery from creep is slow and only partial.

Creosote. (*chem*) A liquid distillation product of coal tar, water-gas tar or wood tar consisting mainly of aromatic hydrocarbons. Creosotes from each source have been used as wood preservatives, but currently, as in the past, coal tar creosote is by far the most important. Its use was patented by John Bethell in 1838 (B.P. 7751) and it is most effective when applied by a full-cell or empty-cell pressure process, although much is applied to fencing and similar structures by brush.

Coal tar creosotes for use in wood preservation are fully defined by several standards e.g. B.S. 144, A.W.P.A. Standard P1. A boiling range of at least 125 Centigrade degrees is usually specified with an initial boiling point of about 200°C. Creosote is often very viscious at ambient temperatures, and before use is usually heated to around 85°C to reduce its viscosity and hence increase its penetration into wood.

Creosote is slightly more dense than water and contains compounds of acidic and basic nature. It does not usually corrode metals, but treated wood cannot be painted easily because of its oiliness. The strong smell of creosote prevents its use in buildings where water-borne or organic solvent preservatives are usually used.

Cross bridging. (*bldg*) Herring-bone strutting in the USA.

Crustacea. (*zoo*) A class of the phylum Arthropoda containing the crayfishes, crabs, shrimps, woodlice, barnacles, waterfleas, and many other diverse forms, the great majority of which are aquatic either in freshwater or marine. The body is segmented as are the generally large number of appendages which are variously modified according to function. Two associated species are well known as causing great damage to wood floating or submerged in the sea. These are the Gribble, *Limnoria lignorum* in the order Isopoda, and *Chelura terebrans* in the order Amphipoda.

Cryptomeria japonica. (*for*) A Japanese softwood timber known as

Sugi but has also been called Japanese Cedar. See *Cedrus*. It is the main softwood species in Japan. Forests denuded in the war. Now being replanted extensively. Good poles.

Cryptotermes. (*ent*) A genus of termites in the KALOTERMITIDAE (drywood) including the following major pests of buildings, *C. brevis* in North, Central and South America, Caribbean, West and South Africa, South Atlantic, Far East and the Pacific, *C. cavifrons* in Central America, northern Caribbean, Bahamas and Bermuda, *C. cynocephalus* in Philippines, Malayan Region and Ceylon, *C. domesticus* in Malayan Region, Far East, Pacific and Central America, *C. dudleyi* in Philippines, Malayan Region, Australia, Central and South America, Caribbean and East Africa, *C. havilandi* in West Africa, South America and Caribbean and *C. pallidus* in Mauritius.

Cuboidal decay. (*myc*) Cuboidal cracking. Decay of wood characterised by the wood cracking into cubes as it loses volume. Decay brought about by the Dry Rot Fungus, *Merulius lacrymans*, shows this feature to an outstanding degree. See Plate 12.

Culture. (*myc*) The cultivation of micro-organisms in prepared nutrient media.

Cupressus. (*for*) A number of species of this genus are grown in East Africa (*Cupressus lusitanica*, *C. macrocarpa*—Monterey Cypress) and elsewhere and are known as East African Cypress, etc., according to origin. See *Chamaecyparis*.

Curculionidae. (*ent*) A family of Coleoptera known as the 'weevils' with about 35,000 species. The largest assemblage of species in any single family of insects, indeed, within the whole animal kingdom. The bark beetles are also sometimes included as the sub-families SCOLYTINAE and PLATYPODINAE. Most weevils can be identified by the enlarged 'snout-like' rostrum and the clubbed and elbowed antennae. In some species the female uses the rostrum as a boring instrument for egg-siting but the function of the rostrum generally is unknown. Although the majority of weevil species are of sombre coloration, some species are of intense brilliance. All the larvae are legless and there is virtually no part of the vegetable kingdom which is immune from the attention of one or more weevil species. *Euophryum confine* and *Pentarthrum huttoni*, usually associated with Wet Rot attacks, are common in South-Eastern England. See separate entries for *Scolytinae* and *Platypodinae*.

Cutaneous. (*tox*) The application of a substance (wood preservative) to the body of an animal by placing it in contact with the surface of the skin or integument.

Cylinder. (*equip*) See *Pressure chamber*.

Cypress. (*for*) See *Cupressus*.

D

Dacrydium cupressinum. (*for*) A softwood species known as Rimu from New Zealand where it was, at one time, the chief house-building timber. It is of good quality rather like New Zealand Kauri but is not durable. Rimu forests are now depleted. Very long rotation, very little replanting. *Pinus radiata* becoming more important.

Dado. (*bldg*) A covering such as panelling on the lower half of a wall in a room but above the level of the skirting.

Daedalia quercina. (*myc*) A basidiomycete fungus attacking hardwoods producing a reddish-brown cubical rot. Sporophore is a thick bracket with elongate pores forming a reticulate pattern. Sometimes found in buildings, usually oak window sills where rain is not quickly drained away.

DDT. (*chem*) $C_{14}H_9Cl_5$ (354·5) The common name approved by BSI for the technical product of which pp'-DDT is the predominant component; the percentage of pp'-DDT should be stated. Its BP name is dicophane; that of the USP is chlorophenothene. pp'-DDT is the common name approved by BSI for 1,1,1-trichloro-2,2-di-(4-chlorophenyl)ethane, also known as dichlorodiphenyl-trichloroethane, hence DDT.

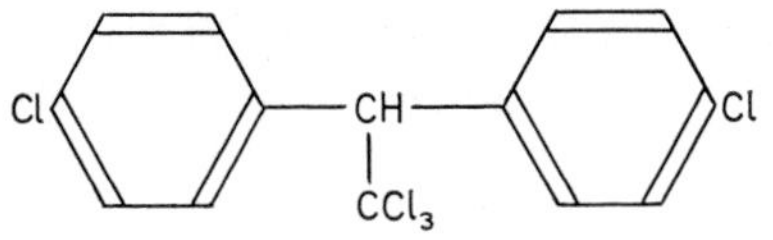

It was introduced, in 1942, by J. R. Geigy S.A. under the trade names 'Gesarol', 'Guesarol' and 'Neocid', with the protection of Swiss P226,180; BP547,871.

DDT is synthesised by the condensation of 1 mole of chloral with 2 moles of monochlorobenzene in the presence of sulphuric acid. The product contains up to 30 per cent of the op'-isomer which, being of insecticidal value, is not usually removed. The pp'-isomer forms colourless crystals of mp 108·5°C, vp $1{\cdot}9 \times 10^{-7}$ mm Hg at 20°C. It is practically insoluble in water, moderately soluble in hydroxylic and polar solvents and in petroleum oils, and readily soluble in most aromatic and chlorinated solvents. The technical product is a waxy solid of indefinite mp and of similar solubility to the pp'-isomer. pp'-DDT is dehydrochlorinated by alkalies or organic bases. Otherwise it is stable being unattacked by acid and alkaline permanganate or by aqueous acids and alkalies, With DDT, dehydrochlorination may proceed at temperatures as low as 50°C.

DDT is a potent non-systemic stomach and contact insecticide of high persistence on solid surfaces. One of the first of the persistent organochlorine insecticides, it found wide favour for increasing the insecticidal action of wood preservatives, especially indoors, and was used widely as a soil treatment against termites.

Its acute oral LD_{50} for male rats is 113 mg/kg, for female rats 118 mg/kg; the acute dermal LD_{50} for female rats is 2,510 mg/kg. Though stored in the body fat and excreted in the milk, seventeen human volunteers ate 35 mg/man/day, about 0·5 mg/kg/day, for twenty-one months without suffering ill-effects.

DDT may be identified by the evolution of HCl when heated above mp or by heating in a 0·5 per cent solution of hydroquinone in N-free H_2SO_4, when a wine-red colour develops.

Product analysis is by total chlorine by the Stepanow method: AOAC Methods: of the labile chlorine: reflux with alcoholic KOH and determine chlorine by the Volhard method: Wichmann, H. J. *et al.*, *J. Ass. off. agric. Chem.*, 1946, **29**: 188. The content of pp'-isomer may be checked from the setting point derived from the time-temperature curve under specified conditions: WHO Tech. Rep., Ser. No. 34, 1951, p. 43; or by crystallisation under prescribed conditions: CPAC method, FAO *Plant Prot. Bull.*, 1961, **9** (8): 153.

Residues may be determined (a) by the Schechter-Haller method: strip with benzene, nitrate residue, extract with ether and treat with anhydrous sodium methoxide, compare blue colour with standards: AOAC Methods. TDE, methoxychlor interfere and special precautions are necessary if much fat is present, see Sergeant, G. A. and Wood, R., *Analyst*, 1959, **84**: 423; (b) in the presence of TDE, by the reaction with xanthydrol and pyridine in KOH: Claborn, H. V., *J. Ass. off. agric. Chem.*, 1946, **29**: 330; (c) by gas chromatography: Coulson, D. M. *et al.*, *J. agric. Fd. Chem.*, 1960, **8**: 399.

Dean and Stark. (*chem*) A method for the determination of moisture in a chemical material by heating under reflux with an organic liquid which is immiscible with water. The carrier liquid distils into a graduated receiver carrying with it the water which then separates to form the lower layer, the excess carrier liquid overflowing from the trap and returning to the still. The carrier liquid may be petroleum spirit, distilling completely between 90° and 160°C, and with a recovery between 5 per cent and 20 per cent at 100°C, as specified in BS3770:1964 (zinc naphthenate and zinc naphthenate concentrates). Alternatievly, for the determination of the moisture content of wood, xylene is often used as the immiscible liquid.

Death Watch Beetle. (*ent*) See *Xestobium rufovillosum.*

Debarking. (*proc*) See *Barking.*

Decay. (*bio, chem, phy*) Deterioration or falling away from a sound condition. In wood preservation this term has sometimes been used to denote fungal decay only, whereas the full meaning must include deterioration by insect attack and other animals, bacteria, physical factors such as radiant energy, and chemical factors, etc.

Density. (*phy*) The weight of unit volume of a substance, usually expressed as lb/ft^3, or as g/cm^3. In the case of wood it is necessary to specify the moisture content as variation in the latter affects its weight and volume.

Deodar. (*for*) See *Cedrus*.

Dermal. (*tox*) Appertaining to the skin. A portal of entry (through the skin) of a toxic substance into the body of an animal.

Dibutyl phthalate. (*chem*) $C_{16}H_{22}O_4$ (278) The trivial name for di-n-butyl phthalate, also known as DBP. It was first used as an insect repellent during the 1939–45 war but had long been in use as a plasticiser.

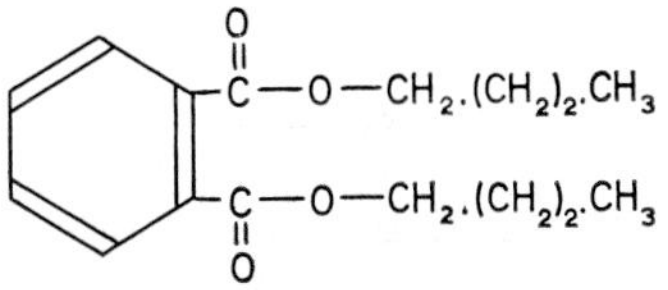

Produced by the esterification of n-butyl alcohol with phthalic acid or phthalic anhydride, it is a colourless to faint yellow viscous liquid of fp about −35°C, bp above 330°C, vp less than 0·01 mm Hg at 20°C, 1·1 mm Hg at 150°C, d_{20}^{20} 1·044, n_D^{20} 1·4926, viscosity 20·3 centistokes at 20°C. Its solubility in water at room temperature is about 400 ppm and it is miscible with ethanol and most organic solvents. It is hydrolysed by alkali.

Dibutyl phthalate has been in use as a anti-blooming agent and auxiliary solvent in formulations of wood preservatives containing pentachlorophenol. It is somewhat less volatile than dimethyl phthalate.

The acute oral LD_{50} for rats is more than 20,000 mg/kg; it is non-poisonous and generally non-irritating to man.

It may be identified by hydrolysis with alcoholic KOH, acidification to precipitate phthalic acid which is identified by the fluorescein reaction. Product analysis is by alkaline hydrolysis and estimation of phthalic acid.

Dichromate. (*chem*) See *Chromate*.

Dieldrin. (*chem*) $C_{12}H_8Cl_6O$ (381) The common name approved by ISO, except in Canada, Denmark and USSR, and by BSI for a material containing not less than 85 per cent of 1,2,3,4,10,10-hexachloro-6, 7-epoxy-1,4,4a,5,6,7,8,8a-octahydro-exo-1,4-endo-5,8-dimethanophthalene. In Canada, dieldrin refers to the pure compound, known in Great Britain as HEOD.

The insecticidal properties of dieldrin were first described by Kearns, C. W. *et al.*, *J. econ. Ent.*, 1949, **42**: 127; it was introduced, in 1948, by J. Hyman & Company under the code number 'Compound 497', trade name 'Octalox' and the protection of USP2,676,547.

Made by the oxidation of aldrin with peracetic or perbenzoic acids: BP692,547, it forms white odourless crystals of mp 175 to 176°C, vp $1{\cdot}78 \times 10^{-7}$ mm Hg at 20°C, d 1·75. It is practically insoluble in water, slightly soluble in petroleum oils, moderately soluble in acetone, soluble in aromatic solvents. The technical product is buff to light brown flakes of setting point not below 95°C.

Dieldrin is stable to alkali, mild acids and to light; it gives no reaction with Grignard reagent and the epoxide ring is unusually stable though reacting with anhydrous HBr to give the bromohydrin. It is compatible with most other pesticides.

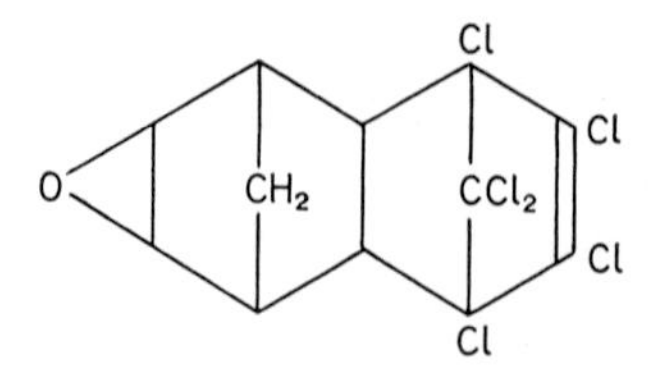

It is a non-systemic and persistent insecticide of high contact and stomach activity to most insects. It has been widely used, particularly in the United States, for soil treatment against termites. In the United Kingdom it was used in wood preservatives of high insecticidal activity and long persistence, but since the publication of the *Review of the Persistent Organo-chlorine Pesticides*, 1964, HMSO, and subsequent reviews, it has been viewed rather less enthusiastically.

Differential Adsorption. (*phy*) The unequal uptake of a preservative fluid by adjacent parts of a wood surface, caused either by variations in grain direction, of extractives content, or by fungal or bacterial action. See *Resinosis, Ponding*.

Diffuse-porous wood. (*bot*) Wood in which the pores, as shown by a transverse section, are of fairly uniform size and distribution throughout a growth ring or showing only a gradual change.

Diffusion treatment. (*proc*) A preservative process in which a water-borne preservative is applied to freshly sawn green timber by steeping or dipping. The timber is then close-piled to allow diffusion of the salts into the cell structure by osmosis.

Dimensional stability. (*phy*) The degree to which timber, either untreated or after treatment with special wood preservatives, retains its dimensions when exposed to cycling conditions of temperature and humidity.

Dimethyl phthalate. (*chem*) $C_{10}H_{10}O_4$ (194) Also known as DMP. It was introduced as an insect repellent during the 1939–45 war, after long use as a plasticiser.

Made by the esterification of phthalic acid or anhydride with methanol, it is a colourless to faintly yellow viscous liquid of bp 282 to 285°C, vp 0·01 mm Hg at 20°C, d_{20}^{20} 1·194, n_D^{20} 1·5168. Its solubility in water at room temperature is 0.43 per cent w/w, it is soluble in petroleum oils, miscible with ethanol, ether and most organic liquids. It is stable though hydrolysed by alkali.

Dimethyl phthalate is an insect repellent used for personal protection from biting insects. Has been in use as an anti-blooming agent and auxiliary solvent in formulations of wood preservatives containing pentachlorophenol. The acute oral LD_{50} for rats is 8,200 mg/kg, the acute dermal LD_{50}, ninety-day exposure, for rats is greater than 4,800 mg/kg; may cause some smarting if applied to the eye or mucous membranes. It had no effect on the growth rate of rats fed, for two years, on a diet containing 2 per cent.

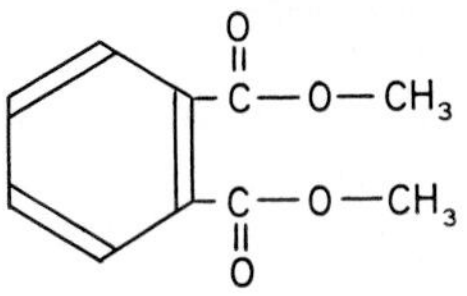

Dipping. (*proc*) A preservative treatment in which the timber is completely immersed in a preservative solution, usually for up to ten minutes.

Diptera. (*ent*) One of the larger orders of insects with probably at least 150,000 species described and undescribed in the world. They are characterised by the possession of one pair of functional wings only, by the second pair of wings being modified into a pair of club-like appendages and by the mouthparts being adapted to sucking and sometimes piercing.

Discoloration. (*chem, bio*) An abnormal change in the natural colour of wood which may be due to weathering, chemicals (including metal contact), the result of fungal attack, etc.

Disinfestation. (*proc*) The removal of an infestation of wood-inhabiting animals from timber.

Door frame. (*bldg*) The surrounding joinery to a door opening to which the door is hinged or carried in some other way. It usually consists of 4 × 3 in. timber.

Door posts. (*bldg*) Door jambs. The two upright posts of a door frame.

Dote or doat. (*for*) A term used to indicate the presence of decay in timber, especially incipient decay, *q.v.*

Double diffusion process. (*proc*) Preservative treatment of timber by steeping with two successive and incompatible water-borne salt solutions.

Double vacuum. (*proc*) A method of applying preservatives which uses only vacuum and atmospheric pressure. The initial vacuum is carefully controlled so that on flooding the treatment chamber with preservative, usually of the organic solvent type, the desired penetration and gross retention can be reached by venting the chamber to the atmosphere. After draining the chamber, a second vacuum is then applied to reduce the retention to the specified value. The process is mainly used for the treatment of building timbers out of ground contact. Less penetration is normally achieved than would be expected with a normal full-cell vacuum-pressure treatment.

Dressed poles. (*proc*) Poles from which all the bark, both outer and inner, has been removed prior to preservative treatment.

Dressed timber. (*proc*) Sawn timber planed on one or more surfaces.

Drier. (*equip*) Equipment such as a kiln for drying materials (e.g. veneers).

Dry rot. (*myc*) Normally used to describe the decay of wood brought about by the Dry Rot fungus, *Merulius lacrymans.* But can be used to describe the decay caused by some other fungi.

Dunkeld larch. (*for*) See *Larix.*

Durability. (*phy/bio*), or **Natural Durability.** The extent to which, without any special treatment, timber maintains its properties (particularly strength) in service. Generally the resistance to biological attack is inferred, in which case the precise class, order or species of both the organism and the wood should be indicated. See *Decay*.

E

Early wood. (*for*) The less dense wood formed during the early period of growth of each annual ring.

East African pencil cedar. (*for*) See *Juniperus*.

Eastern Canadian spruce. (*for*) See *Picea*.

Eave. (*bldg*) That part of a sloping roof, the lowest, which extends beyond the vertical wall.

Electrochemical. (*chem*) A particular type of wood decay usually occurring in boats brought about by the presence of bolts or fastenings of dissimilar metals with sea-water functioning as an electrolyte.

Elm. (*for*) See *Ulmus*.

Elytron. (*ent*) pl. a. The leathery or sclerotised forewing of a beetle serving as a covering for the hindwings when at rest. The elytra usually meet in a straight line down the back.

Emergence hole. (*ent*) A hole made in the surface of wood by the adult stage of an insect biting its way out, after the larval and pupal stages have been passed in the woody tissue.

Emergence period. (*ent*) The interval between the emergence of the first and last adults of an insect species in the course of a seasonal life-cycle.

Emetic. (*tox*) A substance administered orally in order to remove the stomach contents, by way of the mouth.

Empty-cell process. (*proc*) A process of wood preservation by pressure impregnation similar to the full-cell process, with the exception that no initial vacuum is applied. It is designed to secure maximum depth of penetration of the preservative but with limited final retention. See *Lowry process* and *Rueping process*. British Standard 3453:1962.

Emulsion concentrate. (*chem*) A concentrated solution of the active ingredient (insecticide and/or fungicide) in solvents together with emulsifying agents which allow it to be diluted to the required content of active ingredients with water. The emulsion system is a means by which active ingredients such as insecticides and fungicides, not soluble in water, may be applied, using water as a carrier.

End coating or End sealing. (*proc*) A coating applied to the ends of sawn timber during seasoning to prevent a too rapid drying of the ends with consequent checking and splitting.

End grain. (*bot*) The section of wood exposed during cross-cutting, i.e. at right-angles to the grain.

Endrin. (*chem*) $C_{12}H_8Cl_6O$ (381) The common name approved by ISO, except India and South Africa (mendrin), and by BSI for 1,2,3,4,10,

10-hexachloro-6,7-epoxy-1,4,4a,5,6,7,8,8a-octahydro-exo-1,4-exo-5,8-dimethanophthalene; the USA convention requires endoendo. It was introduced in 1951, by J. Hyman & Company under the code number 'Experimental Insecticide 269', protected by USP 2,676,132.

It is made by the epoxidation of isodrin with peracetic or perbenzoic acid. Isodrin is made by the slow reaction of cyclopentadiene with the condensation product of vinyl chloride and hexachlorocyclopentadiene, cf aldrin. It is a white crystalline solid, melting with decomposition above 200°C, vp 2×10^{-7} mm Hg at 25°C. It is practically insoluble in water, sparingly soluble in alcohols and petroleum hydrocarbons, moderately soluble in acetone, benzene. The technical product is a light tan powder of not less than 85 per cent purity, 90 per cent in Canada.

Cl
Cl
O CH_2 CCl_2
Cl
Cl

Endrin is isomeric with dieldrin. It is stable to alkali and acids but strong acids or heating above 200°C cause a rearrangement to a less insecticidal derivative. It is a non-systemic and persistent insecticide used mainly on field crops but has been used extensively for soil treatment in pre-construction use against termites. It is non-phytotoxic at insecticidal concentrations. The acute oral LD_{50} for male rats is 17·5 mg/kg, for female rats 7·5 mg/kg; the acute dermal LD_{50} for female rats is 15 mg/kg.

Engelmann spruce. (*for*) See *Picea*.

Enzyme. (*bio, chem*) An organic catalyst produced by a living cell which makes possible a specific biochemical reaction necessary for a physiological process to take place. A given enzyme molecule is able to catalyze the transformation of from 100 to 3,000,000 substrate molecules per minute at optimum conditions depending on the enzyme involved. Some enzymes act extra-cellularly others only act within the cells in which they are formed (intra-cellular). The degradation of cellulose by fungi takes place outside the cell wall by hydrolysis of its large molecules into water-soluble sugars of low molecular weight. The latter are then taken within the cell and transformed by intra-cellular enzymes to give various products from which the organism derives its energy and substance necessary for its growth.

Ephemeroptera. (*ent*) Mayflies. An order of the class Insecta. About 1,500 species are known of these soft-bodied insects. The antennae are short and the mouthparts vestigial whilst the membraneous wings, the hinder of which are very small, are held vertically together over the body when at rest. The abdomen is terminated by a pair of very long filamentous cerci and often with a similar median filament. The immature stages are aquatic often being found in extremely large numbers in freshwater. The larva of one species, *Povilla adusta*, found in certain lakes in East Africa, as well as a closely-related species found in Asia, damages wooden piling and boats.

Ernobius mollis. (*ent*) A beetle of the family ANOBIIDAE about the same size as *Anobium punctatum* but is light yellowish-brown in colour, and the elytra are soft. The eyes are large and the antennae are thread-like, the individual segments being variable in size but the last three segments are not much stouter than those preceding. It possesses almost a world-wide distribution; indigenous to northern Europe it is especially common in Scandinavia. It has been introduced into northern America and since 1935 has spread to a phenomenal extent in South Africa. It is found also in New Zealand and Australia. The larvae consume the inner bark of a wide variety of coniferous trees but surface tunnels occur also on the outer sapwood, occasionally being much deeper. It is to be found in buildings wherever unbarked softwood is employed. The infestation is identified by the faecal pellets of the larvae, being of two colours, brown when the larva is feeding on bark and creamy yellow when feeding on the sapwood. It is controlled by the removal of the bark which may be present only as a thin strip.

Ethylene dichloride. (*chem*) $CH_2Cl.CH_2Cl$, $C_2H_4Cl_2$ (99) The common name for 1,2-dichloroethane. It was introduced, in 1927, as a component of insecticidal fumigants by Cotton and Roark.

Made by the chlorination of ethylene, it is a colourless liquid with a chloroform-like odour, of bp 83·5°C, mp —36°C, vp 78 mm Hg at 20°C, d_4^{20} 1·2569, n^{20} 1·4443. Its solubility in water at room temperature is 0·43 g/100 ml and it is soluble in most organic solvents. It is inflammable, flash point (Abel-Penchy) 12 to 15°C, lower and upper limits of inflammability in air 6·2 and 15·9 per cent. It is stable, resistant to oxidation and non-corrosive.

Ethylene dichloride is an insecticidal fumigant used mainly for the fumigation of stored products; it has been used for the fumigation of buildings against Dry Wood Termites in the USA.

The acute oral LD_{50} for rats is 670 to 890 mg/kg, for mice 870 to 950 mg/kg, for rabbits 860 to 970 mg/kg. It is as toxic, for exposures of an hour or more, as chloroform or carbon tetrachloride but is less toxic on shorter exposures.

It is usually applied in mixture with carbon tetrachloride in a ratio of 3:1 to reduce the fire hazard. This mixture is used at rates of 8 to 14 lb/1,000 ft^3, or to grain at 3 gal/1,000 bushels.

It may be determined (a) by reaction with magnesium filings in monoethanolamine and estimation of the chloride so produced by the Volhard method, (b) by thermal decomposition and estimation of chloride by the Volhard method.

Eucalyptus. (*for*) A very extensive genus of hardwood trees native to Australia but some species have been widely grown elsewhere. Common timber names for various species or species groups are Australian White Ash, Tasmanian Oak, Southern Blue Gum, Spotted Gum, Jarrah, Saligna Gum, Blue Gum, etc. In addition, there are many hybrids and sub-species. *Eucalyptus marginata*, Jarrah, is one of the strongest timbers and is extremely resistant to decay, especially to marine borers, and thus finds a wide use in timbering in marine work, jetties and wharves. Its resistance to termite attack makes it useful for railway sleepers in tropical countries. This species is very difficult to pressure impregnate, but this is not usually necessary.

European larch. (*for*) See *Larix*.

European spruce. (*for*) See *Picea*.

European whitewood. (*for*) See *Picea*.

Exit hole. (*ent*) See *Emergence hole*.

Exotic. (*for*) Not native to the area. Usually descriptive of tree species introduced from a different country.

Extractives. (*chem*) Those components of wood which are removeable by leaching or extracting with water and organic solvents. This includes gums, resins and most other chemical substances other than lignin, cellulose, hemicelluloses, starch, pectin and ash (inorganic residue).

F

Facing. (*bldg*) Fixed non-structural joinery such as an architrave.

Fagus. (*for*) A genus of north temperate hardwood trees. Principal species are *Fagus sylvatica*, European Beech, *Fagus grandiflora*, American Beech and *Fagus crenata*, Japanese Beech. Liable to attack by fungi in the open but one of the easiest timbers to pressure impregnate. Used for a wide variety of purposes, domestic woodware, turnery, motor-car bodywork, flooring, bentwood furniture. See *Nothofagus*.

Fascia board. (*bldg*) A wide board set vertically on edge and fastened to the wallplate or rafter feet and on to which the gutter is fixed around the eaves.

Fibre saturation point. (*bot*) A theoretical stage in the relationship of water to wood when all the water has been removed from the cell cavities (free water) but none has been removed from the cell walls. In most timbers the fibre saturation point occurs when the cell walls are holding from 25 to 30 per cent of their dry weight of water.

Field test. (*phy/bio*) Any test in which treated specimens are exposed to conditions of natural hazard, as opposed to artificial (e.g. laboratory) conditions; a test of the suitability of treated wood for a particular application. Field tests should always include untreated specimens for comparison with treated ones. See *Accelerated test*, *Graveyard test*.

Fir. (*for*) See *Abies*.

Fireproofing. (*proc*) Applying chemical substances to wood (and other materials), in order that it will not readily ignite or burn.

Fire-retardant. (*proc*) A chemical system applied as a treatment to wood (and other materials) which increases its resistance to spread of flame or retards the rate at which it is destroyed by fire.

Fire-retardant preservative. (*proc*) A chemical system applied as a treatment to wood (and other materials) which increases its resistance to spread of flame or retards the rate at which it is destroyed by fire and in addition imparts preservative properties.

Fistulina hepatica. (*myc*) The Beef-steak Fungus. A basidiomycete usually attacking mature oak where it produces the condition known as 'brown oak'. Sporophore is purplish-red, sessile and tongue-shaped, and exudes blood-like juice when broken. Brown oak has been associated with Death Watch Beetle.

Fixation. (*proc*) The property of certain preservatives of becoming wholly or partially water insoluble after application to timber so that leachability is much reduced.

Flagworm. (*ent*) A popular term sometimes applied to the larvae of ambrosia beetles in the USA.

Flatheaded borers. (*ent*) A popular term given to the larvae of BUPRESTIDAE in the USA on account of the unusually large prothorax. See *Buprestidae.*

Flight. (*bldg*) A series of steps joining one floor or landing to the next floor or landing.

Flight hole. (*ent*) See *Emergence hole.*

Fluoride. (*chem*) Any salt of hydrofluoric acid, HF. Sodium fluoride and ammonium bifluoride are components of commonly used water-borne preservatives, the fluoride ion being toxic to many insects and fungi. The most common fluoride-containing formulations are the fluor-chrome-arsenic-phenol (FCAP) varieties known as Wolman Salts or Osmosar and the fluor-chrome (FC) type known as *Flurasil Ull* or Rentex. In treated wood, part of the fluoride used is insolubilised or fixed by the chromium component.

In central Europe use is also made of silicofluorides, mixed salts of silicic and hydrofluoric acids.

Fomes annosus. (*myc*) A common fungus causing heart rot in coniferous trees but also occurring in dicotyledons and shrubs.

Fontanelle, Fontanel. (*ent*) The pale coloured, shallow depression on the surface of the head of termites into which the frontal pore opens.

Forest Products Research Laboratory or FPRL. (*name*) Set up in 1927 at Princes Risborough, Buckinghamshire, as one of the research stations of the Department of Scientific and Industrial Research, with the object of promoting the most economical use of timber by the wood-using industries. In 1964 it was transferred to the newly formed Ministry of Technology, and since that time it has been increasingly engaged in providing direct assistance to industry. The present staff numbers about 130, excluding industrial staff.

The Laboratory, which covers the whole field of timber utilisation, is organised into three Divisions, each comprising two or three Research Groups. The Structures and Building Division is concerned with the use of timber in buildings. Its main objects are first to improve the efficiency with which timber and wood-based materials are used in situations involving their mechanical strength, by study of the mechanical properties of the materials, the performance of joints, and evolution of design procedures, and secondly, to study the effects of environments on the properties, design and behaviour of wood-based materials and components. Work on the development of efficient drying methods for wood is included.

The work of the Production and Processing Division is designed to improve the efficiency of mechanical processing of timber and deals both with primary conversion (sawmilling) and with subsequent sawing and machining. Work on chemical processing includes plywood and chipboard, wood finishes, improvement in wood properties by combination with plastics, and pulping of home-grown timbers.

The third Division, Anatomy and Preservation, investigates the anatomical features of timbers in relation to their utilisation, and the factors (fungi and insects) causing their deterioration. Biological studies of wood-boring insects are carried out and methods for prevention of infestation or decay and for evaluation of fungicides and insecticides are examined. Work on preservation processes and methods for testing the effectiveness of preservatives is also undertaken. Current projects include the testing of long-term efficiency of wood preservatives against *Anobium punctatum* and practical problems in connection with water repellent formulations.

Publications are classified as Bulletins, Technical Notes, Research Records and Leaflets. In addition, a large number of papers have been published in technical and scientific journals. Books have been published on relevant subjects, as well as a series of booklets on home-grown timbers, and leaflets on the principal woodboring insects and wood-rotting fungi. A list of publications is available on application. Priced publications are purchased from HM Stationery Office and free publications are obtainable direct from the Laboratory, which maintains a mailing list for distribution of new issues.

Foundation pile. (*bldg*) A pile entirely embedded in the ground and capped with masonry.

Frass. (*ent*) Fragments of wood tissue produced by the activity of boring insects, consisting of those fragments torn off but not taken into the mouth (*rejectimenta*) and those fragments which have passed through the gut and been partially digested (*excrementa*).

Fraxinus. (*for*) A genus of north temperate hardwood trees generally known as ash. Three species make up American Ash, *Fraxinus excelsior* is European Ash and *Fraxinus manshurica*, Japanese Ash. This is one of the toughest and most flexible of timbers and finds a wide use in sports goods, motor-body frames, agricultural and garden equipment such as spade and fork handles. It is not resistant to decay, however, so must be preserved if to be used entirely out-of-doors.

Free moisture. (*bot*) Water contained in the cell cavities and intercellular spaces of wood and thought to be held by capillary force only. Cf. *Bound moisture*.

Fruiting body. (*myc*) See *Sporophore*.

Full-cell process. (*proc*) A process of wood preservation by pressure impregnation carried out in four stages as follows:

(i) *Initial vacuum*. The charge of timber is subjected to a vacuum for a period adequate to ensure the subsequent gross volumetric absorption required. This removes some of the air from the wood.

(ii) *Flooding*. The cylinder containing the charge of timber is flooded with preservative solution before the vacuum is released.

(iii) *Pressure period*. The charge of timber is then subjected to a hydraulic pressure of up to 200 lbf/in^2 (14 kgf/cm^2). Pressure is maintained until the gross volumetric absorption is obtained that will ensure both the degree of penetration and the final net dry salt retention required. The period of maximum pressure does not generally exceed four hours.

If the timber is such that the above procedure is inadequate to obtain the desired degree of impregnation, the pressure is maintained until the further absorption in each of any two consecutive half-hour periods is less than 2 per cent of the total absorption up to the commencement of that period.

(iv) *Final vacuum.* At the completion of the pressure period the cylinder is emptied of preservative and a vacuum corresponding to at least 25 in Hg (635 mm Hg) at sea level is established and is maintained only until the net volumetric absorption is sufficient to give the required net dry salt retention. This removes the preservative solution from the wood surface and the process is designed to leave the maximum concentration of the preservative in the treated part of the wood. British Standard 3453:1962 and 4072:1966.

Fumigant. (*chem*) A toxic substance usually an insecticide in the form of a gas, vapour, volatile liquid or solid, used in the atmosphere of a closed chamber, in order to eliminate woodboring insects. The gas enters the insect's body by respiration and diffusion. Fumigants normally possess no residual properties. Buildings or parts of buildings are sometimes covered with gas-tight sheets for fumigation against insects, especially termites. This is known as Tent fumigation, see Plate.

Fumigation. (*chem*) The application of a toxic substance in the form of a gas, vapour, volatile liquid or solid, in the atmosphere of a closed container. The required concentration of gas must be maintained for a specified period and loss of gas by ventilation must be prevented.

Fungal mites. (*zoo*) Small arachnids in the Acarina subsisting on fungal mycelium. If strict attention is not paid to general hygiene in the mycological laboratory, and in the incubators, they may be introduced into the cultures, contaminate them with microfungi, and invalidate results.

Fungi. (*myc*) Plants with true nuclei but no chromatophores. Their nutrition is thus heterotrophic, their mode of life being parasitic (obtaining their nutrition from living organisms) or saprophytic (their nutrition being derived from dead organisms). The wood-rotting fungi belong to this latter group. The cell wall of fungi is more allied to chitin than to cellulose. One of the main groups of the THALLOPHYTA.

Fungicide. (*myc, chem*) In wood preservation, a chemical substance which kills fungi.

Furnos treatment. (*proc*) A preservative treatment for standing poles developed in Sweden, in which visibly decayed wood is removed and

Plate 1: above: Tarpaulin cover being lowered into position to cover spire of St Paul's Cathedral, Wellington, New Zealand, prior to fumigation against *Anobium punctatum*.

Plate 2: right: Tarpaulins being secured at top of spire of St Paul's Cathedral. TV cameramen on platform of hoist in foreground. Fumigation is not in general use against *Anobium punctatum*, but the nature of the wooden spire made this method particularly attractive.

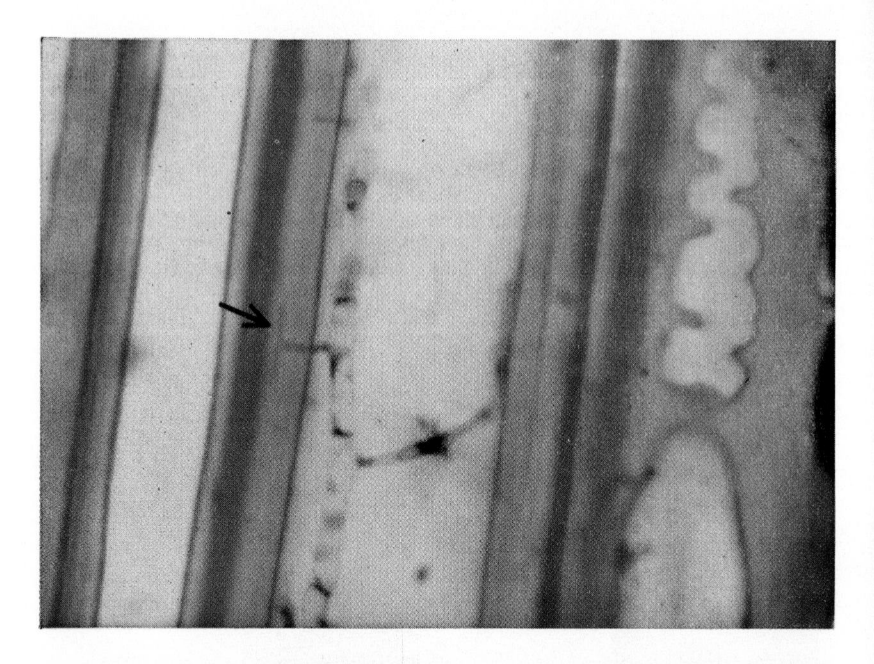

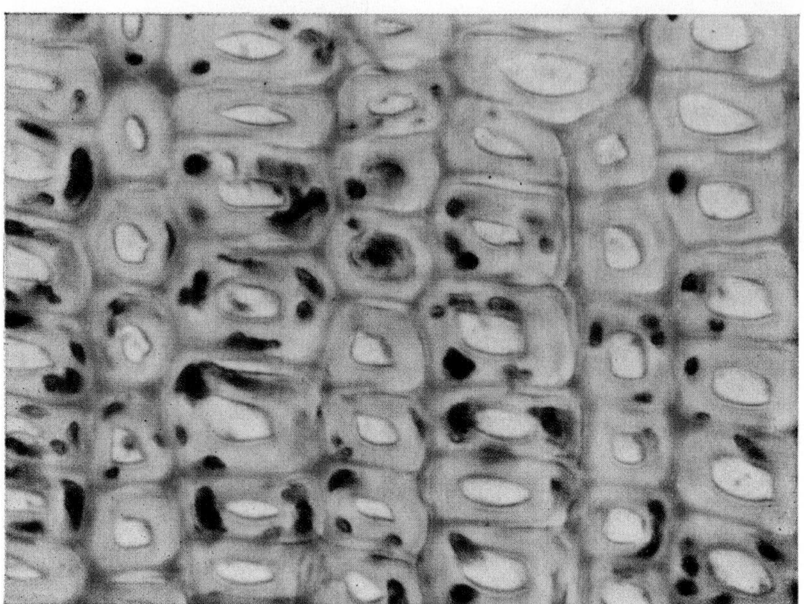

Plate 3: top: Typical T-branching of a soft rot fungus *(Chaetomium globosum)* in the cell wall of *Pinus sylvestris.* × 3,150. John Levy.

Plate 4: below: Transverse section of Scots Pine showing attack by the soft rot fungus *Chaetomium globosum.* × 300. John Levy.

Plate 5: above: Two samples of timber, one decayed by *Coniophora cerebella*—wet rot (above) and one by *Merulius lacrymans*—dry rot (below). Decay of this type is normally associated with dry rot and therefore illustrates the need to take all features into account so that wet rot is not confused with dry rot.

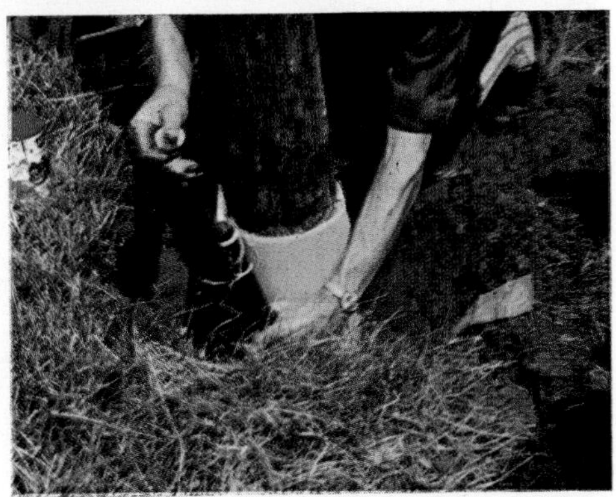

Plate 6: right: Application of a paste-covered bandage to a transmission pole in Ground Line Treatment.

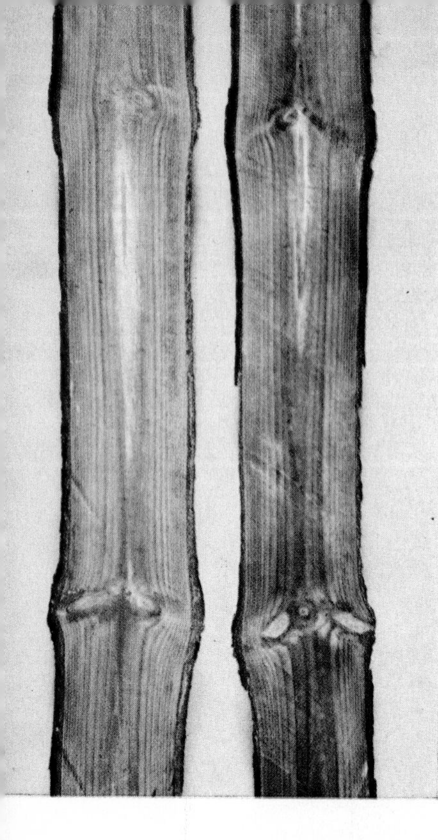

Plate 7: left: Longitudinal section of Scots pine preserved by vacuum cup/sap-displacement using CCA. On left, unstained. On right, sprayed with chrome azurol S reagent.

Plate 8: below: A Kollé flask used in testing the toxicity of wood preservatives to fungi described in BS839: 1961. The wooden blocks are shown enveloped by the fungal mycelium.

Plate 9: opposite, above: unpreserved Scots pine sprayed with benzidine/nitrite reagent in order to differentiate between heart and sapwood. *Below:* Adjacent section treated with Rentex (chromium bifluoride) and sprayed with zirconium alizarin reagent to show penetration into sapwood and heartwood by fluoride.

Plate 10: opposite below: Timber attacked by *Teredo*, the shipworm. Identification of timber attack by *Teredo* is facilitated due to the calcareous material which is used to line the galleries within the wood.

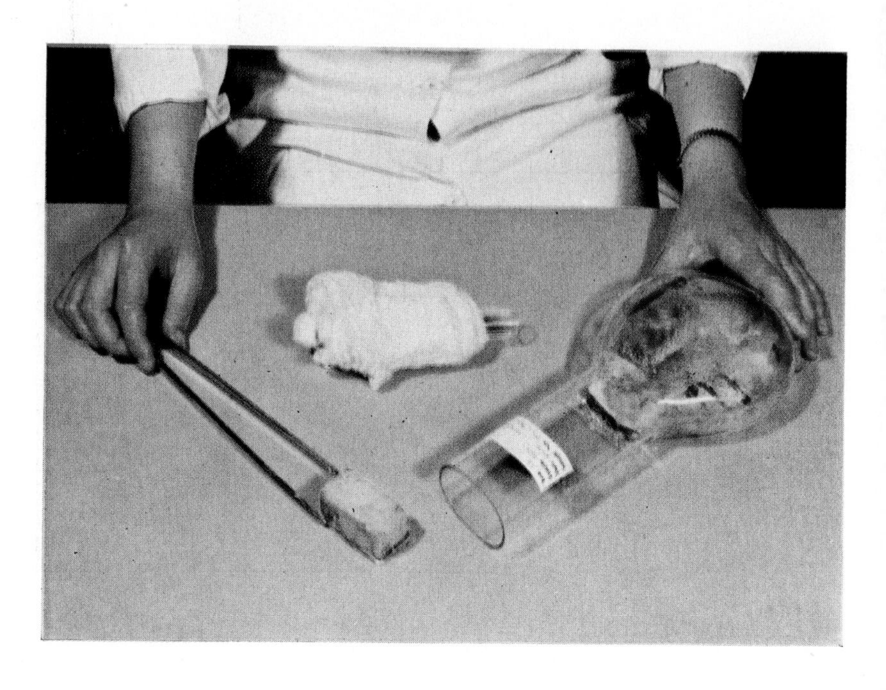

Plate 9

Plate 10

Plate 11: A section of painted joinery decayed by a wet rot fungus. Decay of this type is very common due to extensive use of susceptible sapwood for joinery manufacture.

Plate 12: Extensive mycelium of *Merulius lacrymans* growing under humid conditions. It is in this state that tears are excreted by the fungus, giving rise to the specific name *lacrymans*.

Plate 13: Buprestis aurulenta emerges from Douglas Fir timber where it sometimes spends as long as 26 years in the larval stage, although generally a pest of the standing tree. 12–22 mm in length.

Plate 14: below: The Wharfborer *Nacerdes melanura.* Its larvae are found in wet wood usually attacked by the wet rot fungus, *Coniophora cerebella.* The three longitudinal ridges on each elytron serve to identify it from unrelated but similar species. The Wharfborer has been of particular importance as a pest of wet timber in wooden-hulled sailing ships.

Plate 15:
SCHEMATIC DIAGRAM OF THE FULL-CELL VACUUM-PRESSURE IMPREGNATION PROCESS

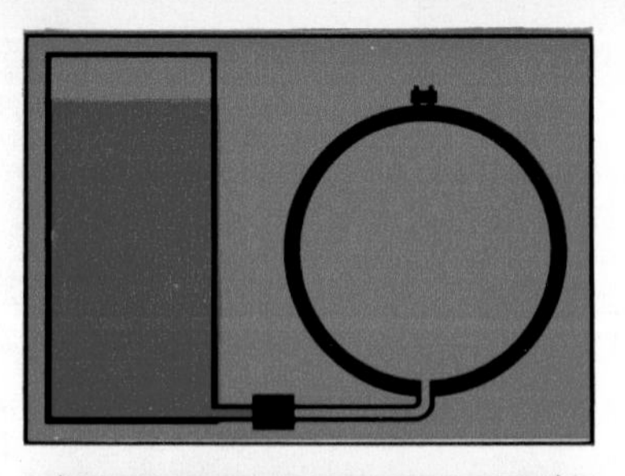

1. Treatment plant before use, with storage tank filled with preservative. Pressure chamber empty.

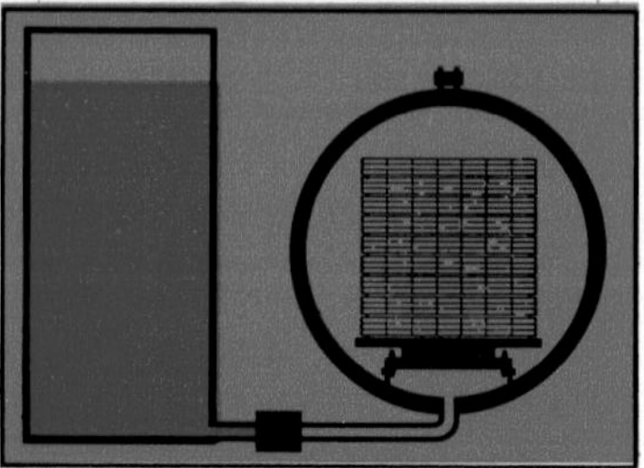

2. A charge of wood has now been moved into the pressure chamber; the chamber is then sealed.

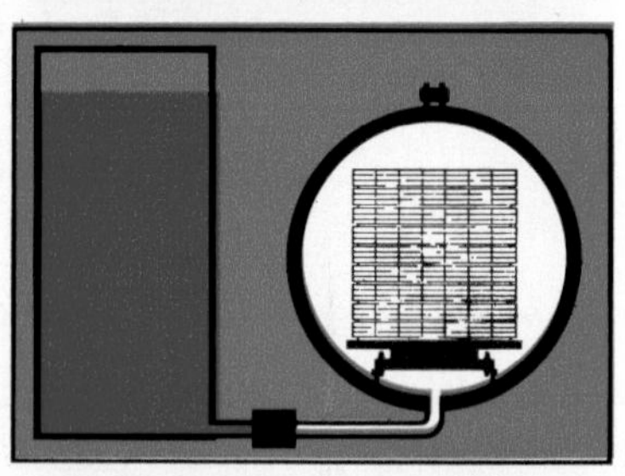

3. The pressure chamber and wood is evacuated by means of the vacuum pump.

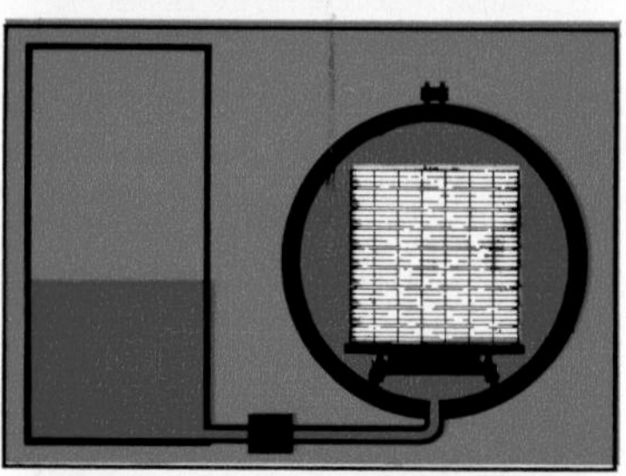

4. Preservative is drawn into the pressure chamber by the vacuum.

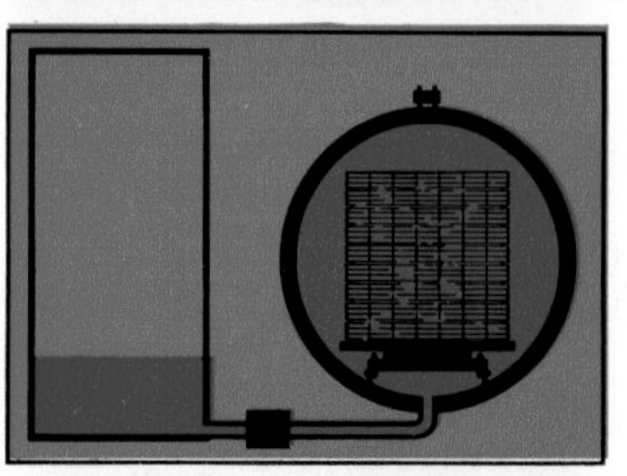

5. Pressure is applied to the preservative in the chamber; preservative is forced deep into the wood.

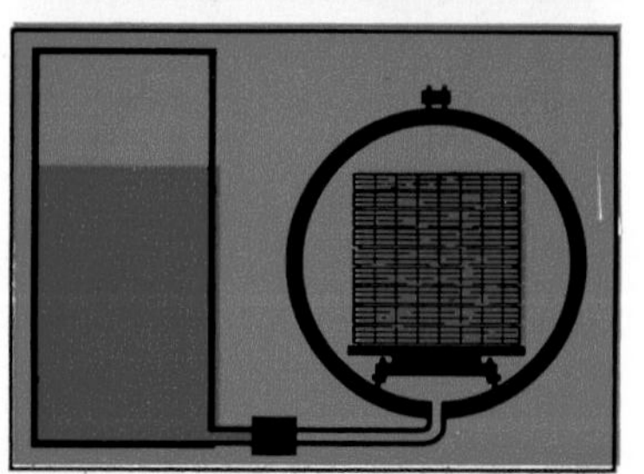

6. Surplus preservative is pumped back to the storage tank, leaving preserved wood ready for removal from the pressure chamber.

the exposed surfaces allowed to dry. These latter are then charred with a blow-torch and then sprayed with creosote.

Gable or gabled end. (*bldg*) That part of the end wall of a building with a pitched roof that extends upwards from eaves level, usually triangular in shape.

Gabled roof. (*bldg*) A pitched roof forming a gable at one end or both ends.

Gambrel roof. (*bldg*) See *Mansard roof.*

Gamma-BHC. (*chem*) $C_6H_6Cl_6$ (291) Gamma-BHC is the common name approved by BSI for the gamma isomer of 1,2,3,4,5,6-hexachlorocyclohexane; the common name of lindane has been approved for this compound by ISO, except Great Britain.

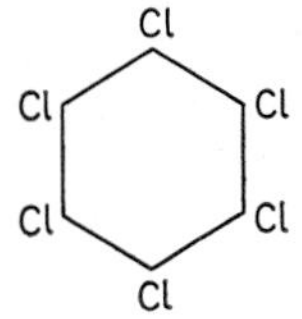

The insecticidal properties of gamma-BHC were discovered in the early 1940s and the compound was introduced by ICI Ltd under the trade name 'Gammexane'.

Cl Cl Cl Cl Cl Cl

Produced by the selective crystallisation of crude BHC—eg, USP 2,502,258, it forms colourless crystals of mp 112·9°C, vp $9{\cdot}4 \times 10^{-6}$ mm Hg at 20°C. Its solubility in water at room temperature is 10 ppm; it is slightly soluble in petroleum oils; soluble in acetone, aromatic and chlorinated hydrocarbons. Lindane is required to contain not less than 99 per cent gamma-BHC and to have a mp of not less than 112°C. It is stable to air, light, heat and carbon dioxide, is not attacked by strong acids but is dehydrochlorinated by alkali.

Gamma-BHC is the main insecticidal component of BHC and has a strong stomach poison action, high contact toxicity and some fumigant activity on a wide range of insects. It is non-phytotoxic at insecticidal concentrations and of lower 'tainting' propensities than BHC. Gamma-BHC has played an important part in increasing the insecticidal properties of organic solvent-type wood preservatives particularly those used against *Anobium punctatum* in buildings. The acute oral LD_{50} for male rats is 88 mg/kg, for female rats 91 mg/kg; the acute dermal LD_{50} for male rats is 1,000 mg/kg, for female rats 900 mg/kg. Rats fed on a diet containing 10 ppm for twelve months suffered no ill-effects.

Product analysis is by a cryoscopic method and a CPAC method. An isotope dilution method has been proposed by Craig.

Residues may be determined by gas-liquid chromatography.

Ganoderma applanatum. (*myc*) A basidiomycete fungus generally attacking living beech trees, but, in addition, a large number of different hardwoods where it produces a heart rot. Sporophore is a large broad flat bracket with brown, silky fibrous flesh. Decayed wood is firstly mottled white, later uniformly white, soft, light and spongy. Limit of fungal advance in wood shown by narrow brown band. Many forms and related species are common throughout the world.

Ganoderma lucidum. (*myc*) A basidiomycete fungus causing a white rot of oak but in America found on conifers. Sporophore is blood-red with varnished appearance and long rounded stalk. Abnormalities, twisted and stick-like, sometimes found indoors.

Gewecke process. (*proc*) A form of sap displacement process where the pole is immersed completely in the preservative solution and vacuum applied at the top end of the pole.

Gill. (*myc*) A plate-like radial lamella on the underside of the pileus and on which the hymenial layer is situated, found in many basidiomycete fungi.

Going. (*bldg*) In a step of a flight of stairs, the horizontal distance measured from the apex of the nosing to the apex of the nosing of the next step.

Grain. (*for*) The appearance on the surface of converted timber caused by the arrangement of the wood cells.

Graveyard tests. (*phy/bio*) A type of field test in which is compared the ability of various preservatives to prevent decay of wooden stakes or fence posts. Generally the posts are installed vertically in the soil outdoors in rows. Similarly-treated posts are often randomly distributed about the test area. A retention range of each preservative is usually applied and the test should contain several replicates of each treatment as well as a number of untreated posts of each wood species exposed.

Green. (*for*) An imprecise term applied to freshly felled or unseasoned timber or to timber still containing free water (or more correctly) liquid sap, in the cell cavities.

Greenheart. (*for*) See *Ocotea rodiaei.*

Green mould. (*myc*) A loose term for a mat of surface-growing green fungi often including species of *Penicillium* and *Trichoderma viride.*

Gribble. (*zoo*) See *Limnoria.*

Gross volumetric absorption. (*proc*) The total volume of preservative solution injected into the charge of timber during a pressure impregnation process, as measured immediately after the release of pressure and before applying a vacuum. It is expressed as gallons of solution per cubic foot of timber.

Ground line treatment. (*proc*) A method of applying preservative to utility poles in service. A paste is formed by incorporating a water-borne preservative in a gelatinous matrix and is applied at ground level in the form of a bandage. It is then protected by a metal or plastic cover. The preservative diffuses into the wood, the bandage preventing loss into the ground. The useful life of the pole is thus extended.

Growth ring. (*for*) A ring of wood, as seen in a transverse section of a trunk or branch, produced in one growing season. Where annual climatic alterations take place, a growth ring may be synonymous with a growing period. In some tropical areas there may be no growth rings shown, or they may correspond to growing periods other than annual ones.

Gun injection. (*proc*) A treatment in which a hollow-toothed tool is used to make an incision in timber and deposit a preservative paste within it. See *Cobra process.*

H

Habitat. (*bio*) The place where an organism occurs naturally. The geographical distribution of an organism. The special locality in which a particular specimen of an organism occurs or the type of habit where an organism occurs.

Handrail. (*bldg*) A guard or rail forming the top of a balustrade.

Hardwood. (*for*) The timber of trees of the class ANGIOSPERMAE. They are mostly broad-leaved. Hardwood is characterised by the possession of vessels in the woody tissue which are absent in the softwoods. The degree of hardness (or softness) is not a critical factor. The terms 'light' or 'soft' hardwoods are in use in some countries.

Heart rot. (*myc, for*) The result of fungal decay of the heartwood of standing trees.

Heartwood. (*for*) The inner core of wood, which in the growing tree has ceased to contain living tissue and in which reserve food materials are absent. It is usually, but not always, differentiated from the sapwood by its darker colour. Generally rather more resistant to the action of wood-destroying organisms than is the sapwood and less easily pene trated by preservatives.

Heel strap. (*bldg*) A U-shaped steel strap which is bolted to the tie-beam of a wooden truss near the wall-plate and passing over the back of a principal rafter, joins it to a tie-beam.

Heptachlor. (*chem*) $C_{10}H_5Cl_7$ (373·5) The common name for 1,4,5,6,7, 10,10-heptachloro-4,7,8,9-tetrahydro-4,7-methyleneindene. It is also known as heptachlorodicyclopentadiene. It was first isolated from technical chlordane, qv, and was introduced, about 1948, by the Velsicol Chemical Corporation, under the code numbers 'E 3314' 'Velsicol 104'.

It may be prepared by (a) the action of sulphuryl chloride on chlordene (see *Chlordane*) in the presence of benzoyl peroxide: BP 618,432; (b) the chlorination of chlordene in the dark in the presence of fullers earth: USP2,576,666.

It is a white crystalline solid with a mild camphor odour, mp 95 to 96°C, vp 3×10^{-4} mm Hg at 25°C. It is practically insoluble in water but its solubility in ethanol is 4·5 g/100 ml and, in kerosene, 18·9 g/100 ml. The technical product contains about 72 per cent heptachlor and 28 per cent of related compounds and is a soft waxy solid of melting range 46 to 74°C, d^9 1·57 to 1·59, viscosity 50 to 75 centipoises at 90°C. It is stable to light, moisture, air and moderate heat and is not readily dehydrochlorinated. It is susceptible to epoxidation and is compatible with most pesticides and fertilisers.

Heptachlor is a non-systemic stomach and contact insecticide with some fumigant action. Has been in wide use in the United States for soil treatment against termites. The acute oral LD_{50} for male rats is 100 mg/kg, for female rats 162 mg/kg; the acute dermal LD_{50} for male rats is 195 mg/kg, for female rats 250 mg/kg. Epoxidation is an important metabolic reaction first observed in dogs and rats by Davidow, and Radomski; the epoxide being highly persistent and biologically active.

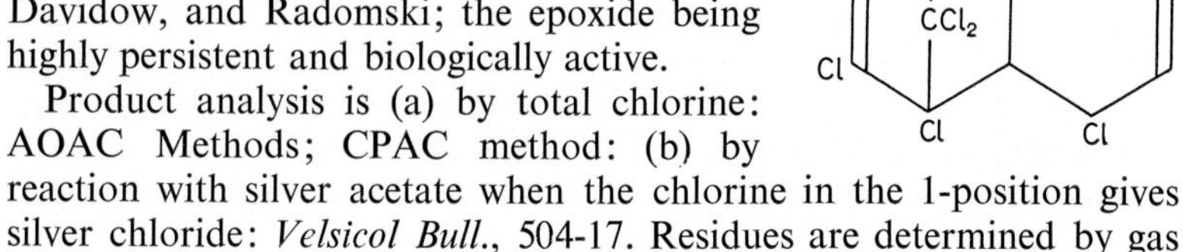

Product analysis is (a) by total chlorine: AOAC Methods; CPAC method: (b) by reaction with silver acetate when the chlorine in the 1-position gives silver chloride: *Velsicol Bull.*, 504-17. Residues are determined by gas chromatography.

Herringbone strutting. (*bldg*) A method of stiffening floor joists at their midspan by fixing light wooden struts from the bottom of each joist to the top of the next, and from the top of each to the bottom of the next, so that they cross in the centre.

Heterotermes. (*ent*) A genus of termites in the RHINOTERMITIDAE (dampwood and subterranean), including the following major pests of buildings. *H. ceylonicus* in Ceylon, *H. convexinotatus* in Central and South America and the Caribbean, *H. philippensis* in the Philippines, Mauritius and Madagascar, and *H. tenuis* in Central and South America and the Caribbean.

Heterothallic. (*myc*) Applied to a number of species of the higher fungi which require a fusion of hyphae from separate spores of different 'sex' before the mycelium can produce a sporophore.

Hip. (*bldg*) The outstanding edge formed by two adjacent roof surfaces where the end of the roof does not terminate with a gable.

Hipped ends. (*bldg*) See *Hipped roof.*

Hipped roof. (*bldg*) A pitched roof of four slopes, the two shorter of which are triangular in shape and are known as the hipped ends.

Honey fungus. (*myc*) See *Armillaria mellea.*

Hoop pine. (*for*) See *Araucaria.*

Host. (*bio*) An organism in which another organism spends part, or the whole, of its existence and fr om which it derives nourishment or protection.

Hot and cold open tank treatment. (*proc*) The application of a wood preservative to seasoned timber by first heating the wood in a tank of preservative solution and then transferring it rapidly to a cold solution of the preservative. It is then kept immersed until the desired absorption of the preservative has been obtained. Alternatively, the timber is immersed in the preservative which is then heated and after an appropriate period, allowed to cool. As the timber cools the preservative is absorbed.

In the case of certain water-soluble preservatives the timber is heated in a tank of hot water before transference to a tank of the cold preservative.

House longhorn beetle. (*ent*) See *Hylotrupes bajulus.*

Hybrid larch. (*for*) See *Larix.*

Hydrocyanic acid. (*chem*) HCN (27) Known also as hydrogen cyanide; the trivial name for its aqueous solution is prussic acid. It was first used as an insecticidal fumigant, in 1886, by Coquillett.

At one time it was prepared, at the site of use, by the action of mineral acids on potassium, later sodium, cyanide; later by the action of moisture on calcium cyanide, qv; more recently by the volatilisation of liquid HCN. It is a colourless gas with an odour, to some people, of almonds; its bp is 26°C at 760 mm Hg. Below 26°C it is a colourless liquid of mp − 14°C, d_4^{20} 0·699. It is soluble in water, ethanol and ether.

Hydrocyanic acid is an insecticidal fumigant for the fumigation of flour mills, stored grain and ships and has been widely used for the fumigation of dwellings against termites, particularly in the USA. It is phytotoxic though usually not at fumigant concentrations. It is powerfully toxic to mammals and concentrations greater than 200 ppm are fatal to man.

It is usually packed in metal containers, with or without an added irritant 'warning' gas; also absorbed on porous material such as kieselguhr or discs of cardboard. It may be detected (a) by the reddening of 1 per cent sodium picrate paper; (b) by the reddening of paper impregnated with a solution containing 2 per cent mercuric chloride, 1 per cent methyl orange and 6·7 per cent glycerol, (c) by the blueing of paper freshly soaked in a freshly prepared mixture of solutions of copper acetate and benzidine: Hydrogen cyanide (Fumigation of Buildings) Statutory Instrument No. 1759, 1951.

It may be determined (a) by conversion, via cyanopyridinium chloride, to glutaconic aldehyde which is condensed with 3-methyl-l-phenyl-5-pyrazolone to give a blue dye measured at 620 mμ, (b) by distilling into alkali, treating with picrate to form potassium isopurpurate and measuring at 500 mμ.

Hylotrupes bajulus. (*ent*) House Longhorn Beetle, a member of the family CERAMBYCIDAE. One of the most widely distributed and most important pests of softwood timber. In Britain it occurs most commonly in a relatively small area of north-west Surrey but elsewhere it causes extensive damage from central Norway to North Africa, and from Portugal to Siberia. In addition, it has been introduced and become established in Australia (where it has since probably been exterminated), South Africa and the United States (where it is known as the Old House Borer). The largest females are up to 25 mm in length whilst the smallest males may be only 7 mm in length. It is greyish-brown to black, covered with greyish hairs and the circular pronotum shows a pair of shining black eye-like marks. On each wing case there are four light grey areas arranged into a pair of sloping transverse bands. The spindle-shaped eggs are laid in two to eight fan-shaped clutches, the total number of eggs laid averaging from 140–200. The larva is cylindrical but flattened somewhat with prominent intersegmental grooves. The prothorax is broad and the head is partly withdrawn into the front end of it. The legs are small but well defined. The larva can easily be heard when it is gnawing in the wood. In Britain the length of larval development is usually from three to six years, but from two to ten years are not uncommon (one of thirty-two years is recorded). It emerges from July to September in Britain.

Hymenium. (*myc*) The spore-bearing tissue of a fungus.

Hymenoptera. (*ent*) One of the larger orders of insects of which about 100,000 species have already been described in the world. They include the wood-wasps (*Siricidae*), wasps, bees and ants, as well as other well-known groups. They are characterised by the possession of two pairs of wings with reduced venation coupled together by rows of hooklets, by the first segment of the abdomen being closely associated with the metathorax and by the occurrence of a constriction between the first and second abdominal segments (with the exception of the *Siricidae* and *Sacoflies*).

Hypha. (*myc*) pl. Hyphae A thread-like element of the vegetative mycelium of a fungus. A single hypha is not usually visible to the naked eye but often the massed hyphae assume macroscopic forms such as rhizomorphs and sporophores.

I

Imago. (*ent*) pl. Imagines. Generally an obsolete term for the adult, reproductive stage of an insect.

Impregnation. (*proc*) The diffusion or saturation of an active substance through a material. In wood preservation, generally used to describe treatments giving a high loading of preservative in the wood as, for instance, in pressure treatments. The term 'pressure impregnation' is often used.

Incipient decay. (*myc*) An early stage of fungal decay in timber. Slight discoloration of the wood can usually be discerned.

Incising. (*proc*) The act of passing timber, usually of species resistant to impregnation such as Douglas Fir, through a machine which cuts or presses a number of holes about ¾ in. in depth into all the surfaces according to a specified pattern. Sawn timbers greater than 3 in. thickness and which are intended for exterior use are usually incised. Deeper and more uniform penetration of the timber is brought about by this means. See British Standards 4072:1966 and 1282:1959, etc.

Increment borer. (*equip*) An auger-like instrument with a hollow bit used to extract a core of wood in order to measure penetration of preservatives, extent of decay, thickness of sapwood, etc. Sometimes called 'hollow borer'.

Incubate. (*bio*) To hatch eggs by means of natural or artificial sustained warmth. To apply a period of artificial sustained warmth, implying temperature and possibly relative humidity control, in order to produce germination or active growth in a micro-organism.

Indigenous. (*bio*) An organism which is native to a specified locality; not imported.

Infection. (*myc*) The act of introducing a harmful micro-organism, either wittingly as an inoculum or unwittingly, into an environment such as a mycological test.

Infection. (*bot*) The invasion of wood, and the subsequent establishment in it of fungi or other plant organisms.

Infestation. (*zoo*) The invasion of wood, and the subsequent establishment in it of insects or other animals.

Inhalation. (*tox*) Intake of a substance by breathing in.

Initial absorption. (*proc*) The amount of preservative absorbed (or taken in) by the wood during the period the cylinder is being filled prior to the application of pressure.

Initial dry weight. (*myc*) Wood blocks or other test pieces are prepared for test by heating in a drying oven at 100 to 105°C for about eighteen hours. They are then weighed aseptically to the nearest 0·01 g and this is known as the initial dry weight. See BS838, Methods of test for Toxicity of Wood Preservatives to Fungi.

Inoculate. (myc) The act of transferring an inoculum under sterile conditions to a suitable nutrient medium.

Inoculum. (*myc*) A small portion of a master culture of a micro-organism or a spore suspension transferred under sterile conditions to a suitable nutrient medium forming a sub-culture.

Insecta. (ent) Insects. A class of the phylum Arthropoda whose members are characterised by the division of the body into head, thorax, and abdomen. A single pair of antennae arise from the head as well as a pair of mandibles and two pairs of maxillae, the second pair joined along the centre. Three pairs of legs as well as, in the great majority of cases, one or two pairs of wings, arise from the thorax. Walking appendages are not borne on the abdomen but the genital aperture is situated at the extremity. In the life cycle metamorphosis is usual. The Insecta is usually classified into twenty-nine orders of which the following contain species known to bore into, or otherwise destroy wood. Ephemeroptera (Mayflies), Isoptera (Termites), Lepidoptera (Butterflies and Moths), Hymenoptera (Wood Wasps, Ants, Bees and Wasps), Coleoptera (Beetles), Diptera (flies).

Insecticidal smoke. (*chem*) See *Smoke generator*.

Insecticide. (*chem*) A chemical substance which kills insects.

In situ. (*proc*) A latin term used for preservative treatment to timber already permanently sited such as a transmission pole or the constructional timber of a building.

Isoptera. (*ent*) Termites. An order of the class Insecta whose 1,800 or so species live in social communities, often of great complexity. The community consists of reproductive forms together with, usually, a large number of wingless, sterile, workers and soldiers of both sexes. The adults (reproductives) bear two pairs of similar wings which are long, narrow and membraneous, and lie flat over the back when at rest. They are cast off from basal sutures shortly after the initial nuptial flight. The tarsi are almost always four-segmented. A genital armature is usually absent or rudimentary in both sexes. Metamorphosis is incomplete, see *Nymph*. Termite nutrition is complex because of the widely differing forms in which the food is taken. But two main types may be distinguished: (a) Feeding on sound or fungal attacked wood, or other vegetable material such as humus, grass, fungi, etc., (b) Feeding on a diet prepared by other members of the colony. This may be taken from the mouth of another termite in which regurgitated intestinal contents is mixed with salivary secretions. Or food may be taken from the anus, after solicitation, which consists of material from the rectal pouch.

J

Japanese cedar. (*for*) See *Cryptomeria japonica.*

Japanese larch. (*for*) See *Larix.*

Jarrah. (*for*) See *Eucalyptus.*

Joggled. (*bldg*) Shaped with a short projection or an indentation usually in order to take the bearing surface of another structural component such as at the base of a King Post.

Juglans. (*for*) A genus of north temperate hardwood trees generally known as Walnut. *Juglans nigra* from eastern USA is American Walnut, *Juglans regia* from Europe, Asia Minor and south west Asia is European Walnut and *Juglans sieboldiana* is Japanese Walnut. This durable and decorative timber has been widely used for high class cabinet work, veneers and especially for gun stocks. Its durable properties (it was used in North America for posts, rails and even for sleepers) were well known before its decorative value was fully appreciated.

Juniperus. (*for*) A genus of softwoods. *Juniperus procera* from East Africa is East African Pencil Cedar and *Juniperus virginiana* from USA is Virginian Pencil Cedar. See *Cedrus.*

K

Kalotermes. (*ent*) A genus of termites in the KALOTERMITIDAE (drywood) including three major pests of buildings, *K. marginipennis* in Central America, *K. minor* in North America and *K. snyderi* in North and Central America and the Caribbean.

Kalotermitidae. (*ent*) A family of drywood termites, ten species of which are serious pests of buildings, especially four species of *Cryptotermes*: *brevis, cynocephalus, dudleyi* and *domesticus.*

Kauri. (*for*) See *Agathis.*

Kickback. (*proc*) The volume of preservative expelled from pressure treated wood when the pressure is released. This is especially important when air has been compressed within the timber. See *Rueping, Lowry processes.*

Kiln-dried. (*proc*) The condition of timber in which the reduction in moisture content has been brought about by exposure in a closed chamber to circulating air, the temperature and humidity of which is suitably varied and controlled.

Kiln-drying schedule. (*proc*) A sequence of time/temperature and humidity conditions to bring about a desired moisture content in timber of a given species and dimensions.

King post. (*bldg*) The central vertical timber of a roof-truss which extends from the ridge at its upper end to the tie-beam at its foot. The latter is often provided with indentations or shoulders in order to form a bearing for two struts. This is referred to as a king post truss.

Knot. (*for*) A transverse section of a branch embedded in the wood of the trunk. They are extensively classified in the timber trade.

Kolle flask. (*myc*) A flat glass vessel of the following dimensions in which fungi are cultured. Diameter of spherical base 135 mm, neck 67 mm, total length 280 mm and depth throughout 50 mm. Two ridges are present at the base of the neck to retain the medium in the base and water in the neck. See BS 838:1961, 'Methods of Test for Toxicity of Wood Preservatives to Fungi'.

Kyanising. (*proc*) A wood preservation treatment involving steeping for eight to ten days in 0·66 per cent corrosive sublimate ($HgCl_2$) solution patented by John Kyan in 1832. Of historic interest only. See *Mercuric Chloride.*

L

Lantern or lantern light. (*bldg*) A glazed frame situated above the general level of a roof.

Larch. (*for*) See *Larix*.

Larix. (*for*) A genus of softwood trees of the northern hemisphere generally known as larch. *Larix decidua* from Europe is European Larch. *Larix eurolepis* from Britain is known as Dunkeld or Hybrid Larch. *Larix leptolepis* (formerly *kaempferi*) from Japan is Japanese Larch. *Larix laricinia* from Canada and USA is Tamarack Larch. *Larix occidentalis* from British Columbia and western USA is Western Larch and *Larix sibirica* from western Siberia and north-eastern Russia is Siberian Larch. *Larix decidua*, *Larix eurolepis* and *Larix leptolepis* are also available from British plantations. The timber is fairly heavy for a softwood. The sapwood is light yellow-brown and the heartwood, well differentiated, is dark red-brown. It is about 60 per cent harder than Baltic Redwood. The timber is used without preservative treatment for many outdoor purposes such as for estate purposes, mining timber, piling, boat building and wooden buildings.

Larva. (*ent*) Immature stage of an insect with a complete metamorphosis. A larva hatches from an egg, and when mature changes into a pupa (qv). Characterised by having a form (and often a habitat) which is totally different to that of the adult. The wings are never visible. Cf *Nymph*.

Larval transfer. (*proc*) A method of testing the efficiency of insecticides against the larval stage of wood-boring insects by transferring larvae of the test insect from rearing blocks or from naturally infested wood to treated standard wood blocks in which holes have been drilled. The subsequent mortality of the transferred larvae after a standard period is observed usually by radiography. By the use of a number of dilutions of the test material its toxic threshold can be estimated. See BS 3651:1963.

Larvicide. (*chem*) A substance which kills larvae. In wood preservation the larvae of woodboring insects is implied.

Late wood. (*for*) The denser wood formed during the later period of growth of each annual ring.

Lauan. (*for*) See *Shorea*.

Lawsons cypress. (*for*) See *Chamaecyparis*.

LD_{50} (*tox*) The median lethal dose. The amount of a substance which kills 50 per cent of a statistically significant group of test animals, ordinarily ten or more. In the case of mammals, the LD_{50} of a substance is usually given in milligrams per kilogram of body weight of the test animals. Other subscripts may be used, indicating the percentage of animals killed by the stated amount of substance.

Leach. (*chem*) To remove water soluble constituents by repeated washing in water. Of importance in determination of the amount of fixation of wood preservative which has occurred. See *Soxhlet*.

Leachate. (*chem*) That which is leached out. In a test of the leachability of a wood preservative, that which has been dissolved out by the water.

Leachability. (*chem*) A measure of the loss of active wood preservative from wood due to repeated washing with water. The natural leachability of preserved wood situated out-of-doors is sometimes implied but generally reference is made to laboratory tests where chemical analysis of the preservative components is carried out before and after a cycle of water washings. See BS 838:1961 with amendment 1965, 'Methods of Test for Toxicity of Wood Preservatives to Fungi'.

Lentinus lepideus. (*myc*) A basidiomycete wood-rotting fungus widespread in Europe and North America. It is resistant to heat and to creosote and is commonly found in railway sleepers, transmission poles and wood-paving blocks containing less than 12 kg of creosote per cubic metre of wood. Wood decayed by the fungus is dark in colour, with cracks along and across the grain with whitish mycelium in the longitudinal cracks, and with strong aromatic odour.

Lenzites sepiaria. (*myc*) A basidiomycete fungus very common in Europe, attacking converted softwood out-of-doors. Uncommon in Britain but often imported. Sporophore is a corky bracket but abnormalities occur in buildings.

Lenzites trabea (*myc*) A basidiomycete fungus sometimes imported into Britain from Europe where it produces a brown cuboidal rot generally in hardwoods out-of-doors, and also in softwoods indoors. Optimum growth temperature is high (35°C).

Lepidoptera. (*ent*) Butterflies and Moths, an order of the class Insecta. They are characterised by the possession of two pairs of membranous wings covered with broad scales and the mouthparts modified into a suctorial proboscis. About 200,000 species have, so far, been described. The larval stage is of great economic importance, devouring the leaves and other parts of herbaceous plants and trees, and a relatively small number of species bore into the woody trunks and branches of trees.

Lethal dose. (*tox*) LD. The amount, or dose, of a substance which will prove fatal to a specified animal or other organism. This is not a precise designation. See LD_{50} and *Minimum lethal dose*.

Leuco base stain. (*chem*) A method for the detection of pentachlorophenol in treated wood. A leuco base dye intermediate, 4,4'-bis-dimethylaminotriphenylmethane has been found useful for the detection of pentachlorophenol in the treated wood. The leuco base in acetone solution is applied to the treated wood and a green colour forms indicating the presence of pentachlorophenol. As little as 0·01 pounds of pentachlorophenol per cubic foot in treated Ponderosa pine sapwood is detected by this method.

Libocedrus. (*for*) The species *bidwillii* is known as New Zealand mountain cedar or Kaikawaka. See *Cedrus*.

Life cycle. (*ent*) The period between fertilisation of an insect egg and the death of the adult individual which proceeds from that egg.

Life history. (*bio*) The continuous story of the activities and duration of each and every stage in the complete life cycle of an organism during one generation or annual series of generations.

Life stage. (*ent*) In insects, those several periods during the life cycle, which are radically different from each other, such as egg, larva and pupa.

Lime. (*for*) See *Tilia*.

Limnoria lignorum. (*zoo*) Gribble, one of the marine borers. A species of Isopoda in the Crustacea causing damage to timber submerged or floating in the sea. It is like a woodlouse, about one-fifth of an inch in length, its tunnels keep close to the surface of the wood, the outer skin of which is continuously washed away by wave action to leave a characteristic 'gribbled' appearance.

Lindane. (*chem*) See *Gamma-BHC*.

Liriodendron tulipifera. (*for*) A hardwood timber species from eastern USA and Canada and known as American Whitewood, Canary Whitewood, Tulip Tree. Not to be confused with the softwood species known as Whitewood.

Loading. (*proc*) The amount of wood preservative absorbed (gross loading) or retained (net loading) by timber during treatment, usually expressed in lb/ft^3 or kg/m^3.

Log. (*for*) A cut section in the round of a felled tree.

Log trap. (*for*) A log left on the forest floor to attract certain species of woodboring insects and subsequently to be destroyed.

Log treatment. (*for*) The spraying of recently felled timber to give protection against wood-destroying insects such as LYCTIDAE and PLATYPODIDAE and/or sap-staining fungi, until the timber is converted.

Longleaf pine. (*for*) Also known as Longleaf Pitch Pine, Longleaf Yellow Pine or American Pitch Pine and consists of *Pinus palustris* and *Pinus elliotti*.

Lowry process. (*proc*) An empty-cell pressure process of wood preservation in which the air is compressed solely by the preservative injected under pressure. Patented in the USA by Lowry in 1906.

Lumber. (*for*) Sawn timber or logs prepared for the sawmill. Used mainly in North America.

Lyctidae. (*ent*) A family of beetles which, together with the closely-allied BOSTRYCHIDAE, are known as Powder Post Beetles. There are about 90 species throughout the world, a number having attained cosmopolitan status. The sapwood of wide-pored hardwoods is attacked when the starch content is high. The larvae resemble those of the ANOBIIDAE but may be distinguished by the possession of a large spiracle on the eighth abdominal segment. The powdered wood is soft and floury contrasting with the rather gritty feel of that from *Anobium punctatum*. The adults are smallish and elongate and usually dark or reddish in colour. The antennae are distinctly clubbed, the club consisting of two segments only.

Lymexylidae. (*ent*) A family of elongate beetles with soft integument. About fifty species of world-wide distribution. The larva possesses an enlarged prothorax and well-developed legs. Considerable damage is caused by the larvae drilling tunnels through, often very hard, woods. British genera are *Lymexylon* and *Hylecoetus*.

M

Macrotermes. (*ent*) A genus of termites in the TERMITIDAE (Subterranean) including two major pests of buildings, *M. bellicosus* in Tropical Africa and *M. natalensis* in Tropical and South Africa.

Mahogany. (*for*) See *Swietenia.*

Malt agar. (*myc*) A nutrient medium made with malt extract and agar and used for the culturing of micro-organisms. See *Agar and Agar-plate test.*

Mammalian toxicity. (*tox*) A measure of the harmful effect produced by a substance on a mammal. The species of mammal and the portal of entry should be specified.

Mandibles. (*ent*) The lateral upper jaws of an insect. In wood-boring beetle larvae they are stout, toothed and generally triangular in shape.

Mansard roof. (*bldg*) A pitched roof each side of which has two slopes, the lower being steeper and the upper being flatter. This type of roof allows of greater utilisation of the attic space and dormer windows are usually present. Known as a gambrel roof in the USA.

Maple. (*for*) See *Acer.*

Marine borers. (*zoo*) Species of Crustacea (eg *Limnoria lignorum*) or of Mollusca (eg *Teredo navalis*) which tunnel into wood floating or submerged in the sea.

Marine environment. (*bio*) A habitat for organisms appertaining to the sea. In wood preservation, wood in contact with sea water either wholly or in part such as in harbour works, sea defences, wooden-hulled vessels and electricity generating stations where sea water is used as a coolant.

Marine pile. (*bldg*) A pile partly embedded in bottom soil and partly exposed to salt water; generally subject to attack by marine organisms.

Martesia. (*zoo*) See *Mollusca.*

Mastotermes. (*ent*) See *Mastotermitidae.*

Mastotermitidae. (*ent*) A family of termites containing one species only *Mastotermes darwiniensis* found only in Australia where it is a serious pest of timber structures in Northern Territory and Queensland.

Mateus test. (*myc*) A method of assessing the degree of fungal decay of wooden test pieces by a measurement of the progressive loss in bending strength.

Maxilla. (*ent*) pl. ae. The second pair of jaws in a mandibulate insect usually bearing segmented palps (maxillary palps).

Mechanical attack. (*phy*) The damage done (to wood) as a consequence of the action of mechanical forces resulting in abrasion, fracture, wear or permanent deformation.

Mercuric chloride. (*chem*) $HgCl_2$ (271·5) Also known as corrosive sublimate; older names are bichloride of mercury, perchloride of mercury. It was first suggested for wood preservation by Kyan in 1832, British Patent 6253. Timber was soaked for eight to ten days in a 0·66 per cent solution and the process was called 'Kyanising'. It was in use up to 1951 for the treatment of spruce and for transmission poles. A number or variations and improvements have been described as follows:

'Mixed Kyanising' (German Patent 290,186 of 1914) also included sodium fluoride.

'Dia-Kyanising' (British Patent 253,041 of 1927) where the timber was first steamheated then dried at 150°C and immediately submerged in mercuric chloride solution.

'Deep-Kyanising' (Czech Patent) 13,012 of 1924) where resin-dissolving substances such as trichloroethylene were added during the steamheating process.

Mercuric chloride may be made by the direct combination of chlorine and mercury or by heating a mixture of mercuric sulphate and sodium chloride and collecting the sublimate. It is a white crystalline powder of mp 277°C which sublimes at about 300°C, vp $1{\cdot}4 \times 10^{-4}$ mm Hg at 35°C, d 5·32. Its solubility in water at 20°C is 6·9 g/100 ml, at 100°C 61·3 g/100 ml. It is soluble in ethanol, ether, pyridine.

It is unstable to alkalies which precipitate mercuric oxychloride; it is decomposed in the presence of organic matter by sunlight to metallic mercury, via mercurous chloride and is readily reduced to mercurous chloride and metallic mercury.

Mercuric chloride is a general poison and is strongly phytotoxic. The acute oral LD_{50} for rats is 1 to 5 mg/kg.

Meranti. (*for*) See *Shorea*.

Merulius lacrymans. (*myc*) The Dry Rot Fungus. One of the most important wood-destroying basidiomycete fungi attacking damp, ill-ventilated woodwork in buildings in northern Europe, and the cause of the decay known as True Dry Rot. The sporophore is plate-like or bracket shaped, soft and leathery but characteristically rusty-red when the spores are mature but with a white margin. Under very damp conditions a white cotton wool-like growth of mycelium may be present but more usually this occurs as a thin silvery grey skin with yellow and lilac tinges. Has ability to form rhizomorphs or water-conducting strands able to penetrate brick walls, etc. For appearance of decayed wood see *Cuboidal decay*. See Plates 5 and 12.

Merulius pinastri. (*myc*) A basidiomycete fungus found only in wet situations where it produces a brown cubical rot. Sporophore is thin and fragile. Pore edge has toothlike projection and sclerotia 1–2 mm diameter may be found on the strands.

Merulius tignicola. (*myc*) A basidiomycete fungus rare in buildings with sporophore thinner and paler than in *M. lacrymans* but with more concentrated attack.

Mesothorax. (*ent*) The second segment of the thorax of an insect, immediately behind the prothorax and bearing the second pair of legs and the first pair of wings or, in the case of beetles, the elytra.

Metathorax. (*ent*) The third segment of the thorax of an insect, immediately behind the mesothorax and bearing the third pair of legs and the second pair of wings.

Methyl bromide. (*chem*) CH_3Br (95) The trivial name for bromomethane. Its insecticidal activity was first reported by Goupil in 1932.

Made by the action of hydrobromic acid on methanol, it is a colourless gas of bp 4·5°C, forming a colourless liquid with an odour resembling chloroform, fp −93°C, d^0 1·73. The specific heat of the liquid at 0°C is 0·12 cal/g. Its solubility in water at 25°C is 1·34 g/100 g and it forms a voluminous crystalline hydrate with ice water. It is soluble in most organic solvents, is stable and non-corrosive and is non-inflammable.

Methyl bromide has high insecticidal and some acaricidal properties. It is used for space fumigation in the control of dry-wood termites and, in some cases, for subterranean termites; also for the fumigation of plants and plant products in stores, mills and ships. It is a soil fumigant used for the control of nematodes, fungi and weeds. It is highly toxic to man, the safe upper limit has been set at 17 ppm, above which concentration, respirators must be worn: UK Home Office Leaflet, 1947. In many countries its use is restricted to trained personnel.

It is packed in glass ampoules (up to 50 ml), in metal cans and cylinders for direct use. Chloropicrin is sometimes added, up to 2 per cent, as a warning gas.

It may be detected (a) by the halide lamp, a non-specific test given by all volatile halides; (b) by decomposition by heat to hydrobromic acid which is detected by test papers impregnated with p-dimethylaminobenzaldehyde or with fluorescein.

For its estimation in air (a) oxidise by a heated platinum wire and estimate bromine (b) convert to inorganic bromide by the action of monoethanolamine.

Microcerotermes. (*ent*) A genus of termites in the TERMITIDAE (Subterranean), including two major pests of buildings, *M. diversus* in Middle East and *M. fuscotibialis* in West Africa.

Micro-fungi. (*myc*) An imprecise term of no taxonomic significance, usually used to include those fungi whose classification is in doubt as no sexual process has been observed, the *fungi imperfecti*, and including also the smaller *Ascomycetes* with small sporophores.

Microtermes. (ent) A genus of termites in the TERMITIDAE (Subterranean) including one major pest of buildings, *M. redenianus* in East Africa.

Minimum lethal dose. (*tox*) MLD. The least of a series of graded doses, or amounts, which will kill one individual of a group of test animals.

Moisture content. (*phy*) The average weight of water in a piece of timber expressed as a percentage of the dry weight of the wood. The method usually employed is described in Appendix F. British Standard 4072: 1966. The method of selection of the sample is specified and drying of the sample, which must not be less than 8 g in weight, is carried out in an oven at a temperature between 95 and 105°C. The sample is dried until any further loss in weight is at a rate not exceeding 0·1 g/24 hours. The moisture content, per cent of dry weight, is then given by:

$$\frac{100\,(w-w_1)}{w_1}$$

Where w = weight, in grams, of sample when wet,
w_1 = weight, in grams, of sample after drying.

Moisture gradient. (*for*) The gradation of water content from the outside to the inside of a piece of timber. Generally this is most marked during seasoning when the outer zone of the timber is relatively dry and the central zone is wetter.

Mollusca. (*zoo*) A phylum of unsegmented invertebrate animals, as a rule possessing an exoskeleton in the form of a shell. The class Pelecypoda, containing the bivalved shell-fish (mussels, cockles, oysters, etc) includes several genera such as *Teredo*, *Martesia*, *Bankia* and *Sphaeroma* which are important destroyers of timber floating or submerged in the sea.

Molluscicide. (*chem*) A substance which kills Mollusca. In wood preservation wood-destroying molluscs such as *Teredo* are implied.

Mottled rot. (*myc*) White mottled rot. A decay of hardwoods such as birch, beech and poplar, caused by the Tinder Fungus, *Fomes fomentarius*.

Mould. (*myc*) Species of fungus growing on the surface of organic material in damp ill-ventilated conditions and usually of a woolly or powdery nature. An imprecise term.

Mushroom. (*myc*) The edible sporophore of the fungus *Psalliota campestris*, but the term has been used loosely to cover the mushroom-like sporophores of a number of basidiomycete species.

Mycelium. (*myc*) pl. a. The collective name for a mass of hyphae. The vegetative thallus of a fungus.

N

Nacerdes melanura. (*ent*) The Wharf-borer. An important pest of wet wood usually in coastal and estuarine areas. The adult beetle is up to 12 mm in length, the elytra are soft, yellowish-brown in colour with black apices. The slender larva is up to 30 mm in length with distinct intersegmental grooves and the head is large and yellowish. Dome-like protrusions on the thoracic and the first two abdominal segments are covered with short spinules. Distributed in many parts of the world, it is thought to have originated in the Great Lakes area of North America and to have been distributed by shipping. See *Oedemeridea. See plate 14.*

Nasutitermes. (*ent*) A genus of termites in the TERMITIDAE (Subterranean) including the following major pests of buildings, *N. ceylonicus* in Ceylon, *N. corniger* in Central America, *N. costalis* in Trinidad, Barbados and Guyana, *N. ephratae* in Central America and the Caribbean and *N. voeltzkowi* in Mauritius.

National Poisons Information Centre. (*tox*) Records of formulations and recommended methods of treatment in the event of accidental poisoning, of a wide range of substances used domestically and in industry, are maintained at local centres as follows:

Guy's Hospital,
London, SE1 (Tel. 01–407 7600).

Department of Forensic Medicine,
University New Buildings,
Teviot Place,
Edinburgh. (Tel. 031–229 2477).

Cardiff Royal Infirmary,
Newport Road,
Cardiff. (Tel. 33101).

Queens University,
Belfast. (Tel. 30503).

Natural durability. (*for*) See *Durability*.

Net dry salt retention. (*proc*) The average weight in pounds of dry preservative per cubic foot of timber in the charge, after the completion of a pressure impregnation treatment.

Net volumetric absorption. (*proc*) The volume of preservative solution remaining in the charge of timber immediately after completion of the entire cycle of pressure impregnation treatment. It is expressed as gallons of solution per cubic foot of timber.

Newel or newel post. (*bldg*) A post in a flight of stairs carrying the ends of the outer string and the handrail and supporting them at a corner.

Non-pressure treatment. (*proc*) A wood preservative treatment in which the preservative is applied at normal atmospheric pressure, eg dip-diffusion, brush, spray, dip, hot and cold open-tank treatment and steeping.

Nosing. (*bldg*) The half-round overhanging edge to a stair tread.

Nothofagus. (*for*) A genus of hardwood trees of a number of species of economic value mainly from Australia, New Zealand and Chile. They are sometimes given the name of beech with qualifications.

Nymph. (*ent*) Immature stage of an insect with an incomplete metamorphosis. It hatches from an egg and changes by a series of moults directly into an adult. Characterised by having a form and habitat very similar to that of the adult. Wings are present as external buds which are not functional until the adult stage is reached. cf *Larva.*

O

Oak. (*for*) See *Quercus.*

Obeche. (*for*) See *Triplochiton scleroxylon.*

Ocotea rodiaei. (*for*) A hardwood of great durability and exceptionally high strength from Guyana, known as Greenheart. It is highly resistant to wood-rotting fungi, termites and marine borers. Very difficult to work but is widely used for lock gates, piers, jetties, shipbuilding and fishing rods.

Odontotermes. (*ent*) A genus of termites in the TERMITIDAE (Subterranean) including the following major pests of buildings, *O. badius* in Tropical and South Africa, *O. ceylonicus* in Ceylon, *O. feae* in India and Burma, *O. latericius* in East, Central and South Africa, *O. obscuriceps* in Ceylon, *O. pauperans* in West Africa and *O. redemanni* in Ceylon.

Oedemeridae. (*ent*) A widely distributed family of elongated, soft-bodied beetles usually of fairly bright or metallic coloration. A characteristic of this family is the three raised longitudinal lines on each elytron. One species, *Nacerdes melanura,* the Wharf-borer, is of important economic significance in many parts of the world as a pest of wet wood. See *Nacerdes melanura.*

Oil-borne. (*chem*) An organic solvent type of preservative whose solvent evaporates slowly, thus largely remaining in the wood after treatment.

Open tank process. (*proc*) See *Hot and cold open tank process.*

Oral. (*tox*) Per os. By the mouth. A portal of entry into the body of an animal.

Organic solvent type. (*chem*) The class of wood preservatives which consist of active components dissolved in any organic liquid other than tar oils or creosotes.

Organochlorine compounds. (*chem*) Hydrocarbon insecticides incorporating chlorine. They act on the nervous system of the insect causing unco-ordination. Similar effects are produced in man and other animals. Most organochlorines accumulate in fatty tissues and are highly persistent in soil. See *DDT*, *Dieldrin*, *Gamma-BHC*, etc. Other organochlorine compounds have enjoyed a wide use as fungicides. See *Pentachlorophenol.*

Organophosphorus compounds. (*chem*) Insecticides derived from phosphoric acid. A large number of compounds belong to this group varying widely in their insecticidal activity and toxicity to man. They act on the nervous system by inhibiting the enzyme cholinesterase.

They are not as persistent as are many of the organochlorine compounds and for this reason have not been generally used in wood preservation.

Orthodichlorobenzene. (*chem*) $C_6H_4Cl_2$ (147) Name approved by the International Union of Chemistry but o-dichlorobenzene is favoured by BSI. Frequently known as ODCB. A liquid varying from light brown to almost water white in colour, with characteristic odour. Flash point 151°F. It mixes readily with most oils but is immiscible with water. Manufactured by the chlorination of benzene under set conditions and then separated by distillation. It possesses useful insecticidal properties attributed largely to its fumigant action, and at one time was extensively used in formulations for eradication of wood-boring insects, now largely superseded. However, it is a good solvent for a variety of insecticides and has good wood penetration properties. Also used as an intermediate in the dyestuffs industry, now probably its main use.

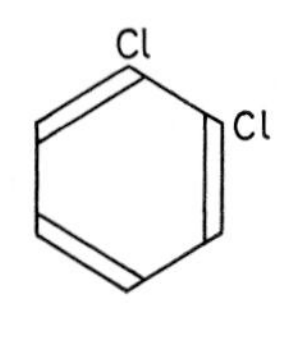

Toxicological information on this compound is limited but intravenous injections with rabbits are stated to have been fatal over the range 326–652 mg/kg. ODCB vapour in air breathed by remedial treatment operators should not be allowed to rise above 50 ppm. There is good reason to believe that the compound has useful emetic properties; a desirable attribute when incorporated into formulations where there is risk of accidental drinking.

Osier. (*for*) See *Salix*.

Osmose process. (*proc*) A preservative treatment in which the freshly felled pole is peeled, brushed with a paste of preservative salts and water, then covered with waterproof paper followed by three months stacking. The salt usually employed consists of sodium fluoride, dinitrophenol, chromates, and sometimes arsenic salts. Patented by Schmittutz, 1932.

Osmosis. (*phy*) The flow of water (or other solvent) through a semi-permeable membrane, ie a membrane which will permit the passage of the solvent but not of the dissolved substances. When solutions of differing solute concentration are separated by such a membrane, water flows from the weaker to the stronger solution.

Oven-dry. (*proc*) A description of wood that has been dried in a ventilated oven at 100°C until there is no further loss in weight.

Ovicide. (*chem*) A substance which kills eggs. In wood preservation the eggs of woodboring insects is usually implied.

Oxine-copper. (*chem*) $Cu(OC_9H_6N)_2$ (351·5) The common name for cupric 8-quinolinolate which is a complex of copper and 8-hydroxyquinoline, the latter long being known by the trivial name oxine. It was introduced for crop protection, in 1946, by Powell.

It is prepared by the precipitation of solutions of soluble copper salts with 8-hydroxyquinoline and is a greenish-yellow, non-volatile crystalline powder, insoluble in water, alcohol and common organic solvents. Chemically it is inert because of the stability of the co-ordination complex.

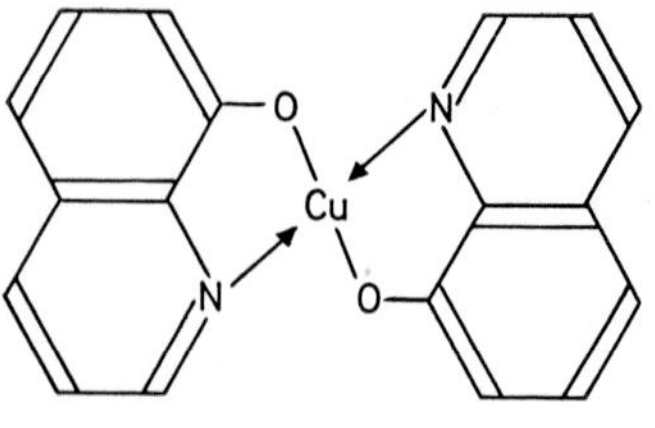

Oxine-copper is a fungicide, of promise for foliage protection, for the mildew-proofing of fabrics and as a wood preservative. It is non-phytotoxic and non-irritating to the skin.

For estimation in timber, the method for copper given in B.S.4072: 1966 can be used.

Oxyacetylene treatment. (*proc*) A preservative treatment for standing poles commonly used in Australia in which an oxyacetylene torch is used for burning out decayed wood and charring the sound wood before application of preservative. Creosote has been generally used for this latter purpose. See *furnos treatment*.

P

Paecilomyces varioti. (*myc*) A species of fungus producing a mould, the spores of which are used in the preparation of a mixed spore suspension in the test for mould and mildew resistance of manufactured building materials. See BS1982:1968.

Parana pine. (*for*) See *Araucaria.*

Parasite. (*bio*) A living organism that subsists directly on or within the living tissues of other organisms.

Parenchyma. (*bot*) Plant tissue consisting of typical, thin-walled, brick-shaped cells which are relatively undifferentiated.

Paste. (*proc*) See *Ground line treatment.*

Parts per million. (*tox*) PPM. The number of parts or units of the substance referred to, contained in one million parts or units (of the same dimension) of the substrate.

Pathogen. (*bio*) A disease-producing micro-organism.

Paxillus panuoides. (*myc*) A basidiomycete fungus occurring from time to time in buildings on damp softwood, but commonly in damp warm mines. Decayed wood first stained vivid yellow then becomes cheesy with deep broad longitudinal cracks but only fine transverse cracks. Sporophore brownish yellow, funnel-shaped with fleshy pileus and branched gills. Hairy mycelium.

Pediculoides. (*zoo*) See *Pyemotes.*

Peeling. (*proc*) The removal of all outer and inner bark of round timber before treatment.

Penetration. (*proc*) The depth to which a preservative enters the wood. Various methods are in use in order to measure this according to the type of preservative used, eg chrome azurol S (Mordant blue 29) is used, as described in BS4072:1966, for copper/chrome/arsenic compositions, and sprayed over a cross section of the redried timber, showing a deep blue colour where the preservative has penetrated. See frontispiece.

Penicillium funiculosum. (*myc*) A species of fungus producing a mould, the spores of which are used in the preparation of a mixed spore suspension in the test for mould and mildew resistance of manufactured building materials. See BS1982:1968.

Peniophora gigantea. (*myc*) A basidiomycete fungus commonly attacking coniferous logs but decay is arrested when timber is converted. Sporophore sometimes produced when fungus is dying. This is irregular waxy crust (spilt candle-grease), creamy white, later brownish or tinged with lilac.

Pentachlorophenol. (*chem*) C_6HCl_5O (266·5) Also known as PCP, penta, penchloral. Trade names include 'Dowicide 7' (The Dow Chemical Company); 'Santophen 20' (The Monsanto Chemical Company).
It was introduced, about 1936, for timber preservation.

Made by the catalytic chlorination of phenol (see USP2,131,259), it forms colourless crystals with a phenolic odour of mp 191°C, vp 0·12 mm Hg at 100°C; it is volatile in steam. Its solubility in water is 20 ppm at 30°C and it is soluble in most organic solvents though of limited solubility in CCl_4 and in paraffinic petroleum oils. Formulations of pentachlorophenol mainly based on the latter require an auxiliary solvent, usually one of high aromatic content, and a non-blooming agent such as dimethylphthalate. The technical product is a dark greyish powder or flakes of mp 187 to 189°C.

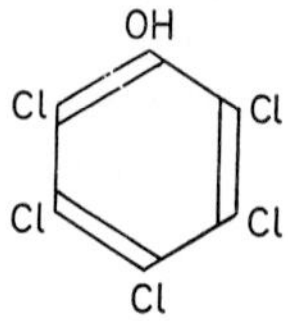

OH
Cl Cl
Cl Cl
Cl

Pentachlorophenol is non-inflammable. Though non-corrosive in the absence of moisture, its oil solutions cause a deterioration of rubber but synthetic rubbers may be used in equipment and protective clothing.

Pentachlorophenol is a fungicide used for the protection of timber from wood-rotting fungi and wood-boring insects: also it is an insecticide used for termite control. When used in buildings where the hazards of attack by wood-boring insects is high, the insecticidal activity of pentachlorophenol solutions is usually increased by the addition of one of the persistent organochlorine compounds. It is strongly phyto-toxic and is used as a pre-harvest defoliant and as a general herbicide.

The acute oral LD_{50} for rats is 210 mg/kg; dogs and rats fed ten to twenty-eight weeks on a diet containing 3·9 to 10 mg/kg/day suffered no fatalities. It irritates mucous membranes and may provoke violent sneezing; the solid and aqueous solutions stronger than 1 per cent cause skin irritation.

Product analysis is by titration with standard alkali: *MAFF Tech. Bull. No. 1*, 1958, p. 64. Residues may be determined (a) by separation from the timber, steam distillation and the formation in chloroform of a brown copper-pyridine-chlorophenol complex. Measure colour intensity at 450 mμ; (b) by oxidation in Parr bomb and determine chloride; (c) by oxidation in benzene solution with HNO_3 to HCl to tetrachloroquinones and compare colour with standards at 450 mμ: *Monsanto Tech. Bull. MeO*—24; (d) form the coloured complex with safranin-O in bicarbonate-buffered solution, extract with chloroform and measure at 520 to 550 mμ: Haskins, W. T., *Anal. Chem.*, 1951, **23**: 1672.

Permeability. (*proc*) In wood preservation, a measure of the ease of penetration by preservatives into a particular timber species.

Permeable. (*proc*) Timbers which can be penetrated completely under pressure without difficulty and can usually be heavily impregnated by the open tank process. See *Resistant*.

Petri-dish. (*bio*) A pair of usually circular flat glass dishes the upper of which, the lid, overlaps the lower. Widely used in micro-biological research work for culturing organisms on nutrient media.

Petroleum oils. (*chem*) Petroleum oils are also known as mineral oils, refined grades have been called white oils. The use of kerosine as an insecticide probably dates from its introduction as an illuminant; oils of higher distillation range came into use about 1922 and highly refined oils about 1924. They are produced by the distillation and refinement of crude mineral oils; those used as pesticides generally distil above 310°C and may be classified as 'light', 'medium' or 'heavy' on the basis of the percentage distilling at 336°C, namely 64 to 79 per cent, 40 to 49 per cent, 10 to 25 per cent, respectively. Viscosity and density vary according to the geographical area from which the crude oils came; but oils come within the range 3 to 11 centistokes, determined at 38°C. For practical purposes, they may be regarded as noticeably thicker than water but still readily pourable. Density rarely exceeds 0·92 at 15·5°C (see *Tar oils*). Flash point normally between 93 and 177°C (closed cup).

Petroleum oils consist largely of aliphatic hydrocarbons, both saturated and unsaturated, the contents of the latter being reduced by refinement. They enjoy a wide use as solvents in organic solvent wood preservatives (type OS), and are relatively harmless to mammals.

Phellinus contiguus. (*myc*) A basidiomycete fungus commonly decaying timber in buildings in New Zealand, but sometimes reported in Britain causing a stringy white rot in window sills and frames, usually of oak, but sometimes softwood.

Phellinus megaloporus. (*myc*) A basidiomycete fungus, also known as *P. cryptarum*, causing a serious rot in oak and chestnut indoors where persistently damp, warm and dark; usually in large-section timbers. Sporophore is plate-like, woody buff to rich dark-brown, pores about 0·25 mm diameter and often stratified. Active mycelium is matted, yellow to red-brown crust exuding a yellow-brown liquid. White stringy rot finally produced which is non-powdery. Decayed wood very light. This fungus causes the most rapid decay of oak in buildings, but mycelium does not travel over inert material.

2-Phenylphenol. (*chem*) $C_{12}H_{10}O$ (170) Also known as 2-hydroxydiphenyl, o-phenylphenol, orthoxenol. It was introduced in 1936 by Tompkins, R. G., 1936, for the treatment of fruit wraps to prevent rot.

It is recovered from the reaction products of NaOH and monochlorobenzene and forms white to pinkish crystals of a mild odour, mp 57°C, bp 286°C, volatile in steam, d 1·217. Its solubility in water at 25°C is 0·07 g/100 g, it is soluble in most organic solvents.

It forms salts of which those of the alkali metals are water-soluble; the sodium salt crystallises with 4 moles of water of crystallisation and is soluble to about 110 g/100 g water, giving a solution of pH 12·0 to 13·5 at 35°C.

2-phenylphenol is a powerful disinfectant and fungicide finding use for the impregnation of fruit wraps and for the disinfection of seed boxes, also for a variety of uses as a wood preservative.

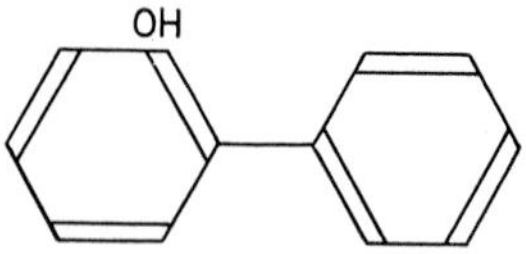

The acute oral LD_{50} for white rats is 2,480 mg/kg; there were no adverse effects on rats fed two years at a dietary level of 0·2 per cent.

It may cause skin irritation.

Phloem. (*bot*) Tissue derived from the cambium and located to the outside of it, forming the inner bark or bast. See *Bark*.

Physical attack. (*phy*) The damage done (to wood) due to physical influences, the most important of which is heat. Some authors include moisture as a physical influence, an excess of which can cause wood to warp and crack. This, however, should more correctly be included under 'Chemical attack'.

Phytotoxic. (*tox*) The property of being harmful to plants.

Phytotoxicity. (*tox*) A measure of the harmful nature of a substance to plant-life.

pH Value. (*chem*) A measure of the hydrogen-ion concentration, and hence, of the acidity or alkalinity of a solution, expressed on a scale ranging from nought for a solution containing 1 gram-ion of hydrogen ions per litre, corresponding to extreme acidity, to fourteen for a solution containing 1 gram-ion of hydroxyl ions per litre.

Picea. (*for*) A genus of softwood trees widespread throughout the northern hemisphere and generally known as spruce. *Picea abies* (was *Picea excelsa*) from northern and central Europe is European spruce, but is also known as whitewood with various qualifying words generally of origin. Whitewood from central and southern Europe also contains *Abies alba*. Confusion may arise in that the hardwood *Liriodendron tulipifera* is known as American Whitewood. *Picea engelmannii* from Alberta, British Columbia and western USA is Engelmann Spruce. *Picea glauca* from the same regions is Western White Spruce. Various species but principally *Picea glauca* from eastern and northern Canada and eastern USA are Eastern Canadian Spruce. *Picea sitchensis* from western Canada and western USA is Sitka Spruce. Sitka Spruce and European Spruce are available plantation grown in the British Isles. The timber of all species is somewhat similar being almost white and slightly lustrous with indistinct heart and sapwood differentiation but whereas good strength properties are shown by Sitka Spruce, the other species can be compared with Baltic Redwood. It is resilient, however, and is much used for ladder sides, cars and masts. Sitka Spruce has very little odour and can be used for food containers. Spruce is greatly used for paper pulp. All species lack durability out of doors but are difficult to impregnate with preservatives.

Pile. (*bldg*) In general, a timber, usually round, embedded wholly or partly in surface or under-water soil, as a support for a superstructure such as a bridge, building, wharf, etc.

Pileus. (*myc*) The umbrella-shaped cap of the sporophore of many basidiomycete fungi.

Piling. (*for*) Placing timber in piles or stacks for storage, seasoning, or preservative diffusion. In 'close piling' the timber is stacked without air spaces between adjacent timbers.

Pinhole. (*ent*) A hole less than 1·5 mm diameter in timber caused by ambrosia beetle. It is characterised by dark staining, absence of bore-dust and penetration of the timber mainly across the grain and in a straight line.

Pinhole borer. (*ent*) See *Ambrosia beetles.*

Pinus. (*for*) An extensive genus of softwood trees preponderantly of the northern hemisphere in origin. The standard name for all species is pine suitably qualified except that *Pinus sylvestris* imported from continental Europe is traditionally known as Redwood but it should be noted that this name is also applied to several softwood species other than the genus *Pinus. Pinus banksiana* from eastern and central Canada and northern USA is Jack Pine. *Pinus caribbaea* and *Pinus oocarpa* from Central America, Cuba and the Bahamas is Caribbean Pitch Pine. *Pinus cembra* and *Pinus koraiensis* from Siberia and Manchukuo is Siberian Yellow Pine. *Pinus contorta* from British Columbia, Alberta and western USA is Lodgepole Pine. *Pinus lambertiana* from Oregon and California is Sugar Pine. *Pinus monticola* from western Canada and western USA is Western White Pine. *Pinus niara* from south-eastern Europe is Australia Pine. *Pinus nigra* var. *calabrica* from south-eastern Europe is Corsican Pine. *Pinus palustris* and *Pinus elliottii* from southern USA is American Pitch Pine. *Pinus pinaster* from southern and south-western Europe is Maritime Pine. *Pinus ponderosa* from British Columbia and western USA is Ponderosa Pine (in USA this includes also timber from *Pinus jeffreyi*). *Pinus radiata* (was *Pinus insignis*) originating in California and now from Australia, New Zealand and South Africa is Radiata Pine. *Pinus resinosa* from eastern Canada and northern USA is Canadian Red Pine. *Pinus strobus* from eastern Canada and eastern USA is Yellow Pine. *Pinus sylvestris* from northern Europe and western Siberia is Redwood or Scots Pine. This latter timber has traditionally been referred to by many other names including red deal, yellow deal and with countries of origin as well as Fir, Norway Fir, Red Pine, Scots Fir, etc.

Pipe rot. (*myc*) A decay of oak heartwood caused by the fungus *Stereum gausapatum* and characterised by the presence of dark-brown streaks in the centre of which there are completely decayed yellowish bands.

Pitched roof. (*bldg*) A roof of two or more slopes meeting at a ridge at the apex.

Platypodidae. (*ent*) See *Platypodinae.*

Platypodinae. (*ent*) A sub-family of the CURCULIONIDAE, the weevils, formerly a separate family, PLATYPODIDAE. They can be considered together with the SCOLYTINAE as both sub-families contain Ambrosia and Bark beetles. See *Scolytinae.*

Plywood. (*proc*) A board made up of plies glued together in such a way that the grain direction alternates in order to improve strength and movement properties. Plywood faced with a decorative wood veneer is known as 'veneered plywood' whilst that faced with a material other than wood, such as plastic or metal, is known as 'faced plywood'. Preservative treatment of plywood is known as pre-treatment when the plies are treated before glueing and post-treatment when treatment takes place after glueing.

Pole bandaging. (*proc*) An *in-situ* process for the preservative treatment of poles in which the surrounding earth is removed from around the ground-line of the pole and concentrated preservative salts applied in the form of a bandage wrapped around the pole. It may then be protected with metal or plastic sheet and the soil replaced. The salts diffuse into the wet wood. See *Ground line treatment.*

Pole plate. (*bldg*) a horizontal beam resting on, and perpendicular to, the tie-beams or principal rafters of a roof truss, supporting the feet of the common rafters.

Polyporus fumosus. (*myc*) A basidiomycete fungus attacking hardwoods out-of-doors with which Death Watch beetle has been associated.

Polyporus sulphureus. (*myc*) A basidiomycete fungus usually decaying old oak trees and other hardwoods but can remain living in large-scale oak timbers, and accidental water leakage can start active growth. Normal sporophore is yellow finger-like whorl but indoors often antler-like. Cuboidal cracking occurs and skin-like mycelium often present.

Polystictus versicolor. (*myc*) A very common basidiomycete wood-rotting fungus in all temperate countries attacking the sapwood of almost all hardwoods, especially when in ground contact such as fence posts and logs lying on the ground. In less durable species such as beech the wood can be completely decayed. The sporophore is a thin tough bracket with upper surface velvety and zoned in greys and browns, and with lower surface cream and pitted with pores. First indications of decay in ash or beech are white flecks, the wood then becomes paler and lighter until practically white and extremely light in weight.

Ponding. (*for*) The practice of storing recently felled trees by floating them in lagoons, rivers or ponds (log-ponds). In this way the trees are prevented from spoilage by insects, fungi and splitting, until needed for sawing or chip/pulp production. The practice is common in Canada, N. America, Finland and Sweden, but is said to be on the decline. Although preventing other forms of degradation, the ponding of timber in still waters is believed responsible for allowing bacterial action in the sapwood, resulting in the seasoned wood being unusually absorbent. See *Differential adsorption.*

Poplar. (*for*) See *Populus.*

Populus. (*for*) A genus of hardwood north temperate trees generally known as poplar with qualifying name but *Populus deltoides* is Eastern Cottonwood, *Populus trichocarpa* is Black Cottonwood, *Populus tremula* is European Aspen and *Populus tremuloides* is Canadian Aspen. The timber generally tends to be woolly and difficult to saw but is useful for many purposes especially for matches (slow burning, burns to splint and not ash) and matchboxes.

Pores. (*bot*) Vessels, vascular or water-containing elements of hardwoods, usually easily visible in transverse sections.

Poria. (*myc*) A genus of basidiomycete fungi, several species of which decay wood in buildings, in mines, and out of doors. They are identified by the sporophore being flat, generally white, and pitted with numerous pores which produce colourless spores. The appearance of the decay is similar to that produced by *Merulius*, but the mycelium and strands always remain white and the latter are flexible when dry. *Poria* does not extend in the wood to any extent beyond the areas of dampness and its subsequent growth is inhibited by ventilation and drying. The principal species in Britain is *vaillantii*, but there has been much confusion in nomenclature, especially with *vaporaria*. The common species in America is *Poria incrassata*.

Poria medulla-panis. (*myc*) A basidiomycete fungus causing a decay of oak in buildings, indistinguishable from that caused by *Phellinus megaloporus*. Sporophore is tough, whitish or light buff and pores measure four or five to the millimetre.

Poria monticola. (*myc*) A basidiomycete fungus with a much confused nomenclature in its history. Does not occur under natural conditions in Britain, but is imported from North America in Douglas Fir and Sitka Spruce. If timber is used in dry well-ventilated position fungus soon dies out, but if in damp ill-ventilated situation, an extensive brown crumbling rot can occur.

Poria vaillantii. (*myc*) The White Pore Fungus. A basidiomycete with a confused history of nomenclature. Has often been referred to as *P. vaporaria*. It occurs in very damp conditions such as in mines, but sometimes in buildings where it flourishes up to a temperature of 36°C. Production of a sporophore is rare in buildings, it is an irregular white plate covered by pores, tubes of which may vary in length from one-sixteenth to half an inch. Only softwoods are attacked and fern-like mycelium does not show lilac and yellow coloration as in *Merulius lacrymans*. Rhizomorphs are smaller than in latter species and remain flexible when dry. Cuboidal cracking also not so pronounced.

Poria vaporaria. (*myc*) See *Poria vaillantii*.

Port Orford cedar. (*for*) See *Chamaecyparis q.v.*

Potentiation. (*chem*) Equivalent to Synergism.

Powder-post beetles. (*ent*) See *Lyctidae* and *Bostrichidae*.

Powell treatment. (*proc*) A diffusion process in which timber is steeped in a sugar solution containing arsenic and sometimes other chemicals. Historical interest only.

Predaceous. (*zoo*) Predatory.

Predator. (*zoo*) An animal that preys on others.

Prefabrication. (*proc*) The cutting, machining to final shape and dimensions, and in some cases the fitting together of timber before preservative treatment. In this way, exposure of untreated surfaces after treatment is prevented.

Preheating. (*proc*) The steeping of timber in hot preservative or treatment with steam before a pressure process is applied in order to increase its permeability.

Preservation (of wood). (*phy*) Protection of wood from influences of a biological, physical, mechanical or chemical nature. A chemical treatment is usually inferred whereby substances are incorporated into or on to the wood to make it resist such hazards for a longer time than would otherwise be the case.

Pressure chamber. (*equip*) The retort, cylinder or autoclave in which wood is placed for any treatment process involving a differential pressure; a chamber capable of safely withstanding pressure differentials.

Pressure process. (*proc*) Any treatment in which the preservative is forced into the timber by pressure applied in a closed vessel, autoclave or pressure cylinder.

Pre-steaming. (*proc*) See *Steaming.*

Principal rafter. (*bldg*) The sloping member of a roof truss that carries purlins

Pronotum. (*ent*) The upper or dorsal surface of the prothorax of an insect.

Propylene oxide. (*myc*) 1:2-epoxypropane. A volatile sterilant used for sterilising test pieces when heating is not admissible. It is used at the rate of 1 ml per litre of space or sufficient to saturate the atmosphere of the container.

Protective threshold. (*biol*) See *Total inhibition point.*

Prothorax. (*ent*) The first segment of the thorax of an insect, immediately behind the head and bearing the first pair of legs.

Psammotermes. (*ent*) A genus of termites in the *Rhinotermitidae* (dampwood and subterranean), including one major pest of buildings, *P. hybostoma* in North Africa.

Ptilinus pectinicornis. (*ent*) A beetle in the family ANOBIIDAE which attacks a number of hardwood species in Europe. Beech, ash, maple, sycamore, oak, alder, plane and elm, have been recorded as supporting infestations. In Germany it has been stated to be second in importance only to *Anobium punctatum* as a pest of drywood, but in Britain it must be placed amongst species of lesser importance. The adult beetle is elongate and cylindrical with a globular pronotum the latter darker in colour than the elytra which are more reddish. The large, strongly pectinate antennae of the male are quite extraordinary, whilst those of the female are serrate. The female bores a brood chamber at right-angles to the grain of the wood or she may bore a tunnel leading from a pupal chamber. Here she lays her long thin-shelled eggs directly into the lumen of vessels and afterwards dies with the hard head and pronotum forming a closure to the egg-laying site. The first instar of the larval stage is remarkable in that the body is extraordinarily long and thin.

Puddling treatment. (*proc*) An *in-situ* treatment for poles and posts in which the soil surrounding the timber is saturated with preservative.

Pupa. (*ent*) The third, resting or quiescent stage, following the larval stage, of an insect with a complete metamorphosis. Chrysalis. Often enclosed in a cocoon made by the larva.

Purlin. (*bldg*) A horizontal beam connecting and lying on top of rafters or trusses and at right-angles to them. It lies between the ridge and the wall plate, and it carries the common rafters if they are present.

Pyemotes ventricosus. (*zoo*) A parasitic mite frequently present in large numbers in wood infested by *Anobium punctatum*, feeding on the beetle larvae. Their presence is often a problem in the laboratory, but a 2 per cent solution of beta-naphthol in 97 per cent ethanol is effective in controlling them.

Q

Queen post. (*bldg*) One of two posts nearest the centre in a roof truss when a king post is absent.

Queen post truss. (*bldg*) A truss in which the central vertical timber (king post) is absent but two vertical timbers are present, one on each side of the centre (queen posts). Usually used for larger spans than the king post truss.

Quercus. (*for*) A genus of north temperate hardwood trees generally known as oak with qualifying name. European Oak consists of the two species *Quercus robur* (sometimes called Common or English Oak) and *Quercus petraea* (sometimes called Durmast or Sessile Oak), and hybrids are common. American Red Oak is a complex of species from eastern USA and Canada, as also is American White Oak. Japanese Oak consists principally of *Quercus mongolica*. Holm Oak is *Quercus ilex* from Europe, and Turkey Oak is *Quercus cerris*. Oak is hard and tough and one of the most useful of all timbers. The heartwood is extremely resistant to decay but very old timbering in buildings in England and Wales often shows attack by Death Watch beetle, *Xestobium rufovillosum* and Common Furniture beetle, *Anobium punctatum*.

Quick-lock door. (*equip*) A variety of pressure-tight door used at one or both ends of a pressure treatment chamber. It is usually either manually or hydraulically locked in a matter of seconds, and has largely superseded the earlier bolt-on doors which were cumbersome and slow in use.

R

Rabbet. (*bldg*) A rectangular recess cut into the edge of timber. Also called Rebate or, in Scotland, Check.

Ray. (*bot*) A thin ribbon of soft tissue radially disposed in the wood. The medullary ray has its origin in the pith or medulla. In certain timber species it may impart 'figure', e.g. oak.

Rebate. (*bldg*) See *Rabbet.*

Red heart. (*for*) A reddish coloration in the heartwood of softwoods and a few hardwoods. In the United Kingdom, a firm, incipient stage of decay caused by species of *Fomes* in softwood is known as Firm Red Heart. Several fungi attacking living trees produce reddish discolorations and in other countries the term may be applied to the coloration produced by fungi other than species of *Fomes.*

Red ring rot. (*myc*) A decay, usually of standing conifers, characterised firstly by a pronounced purplish discoloration of the heartwood in which resin may be heavily infiltrated, often termed 'redheart' and finally by the presence of numerous white, fibrous, pointed pockets. The latter may be so numerous that they merge to form a mass of white fibres. The causative fungus is *Fomes* (*Trametes*) *pini.*

Refractory (*proc*) In wood preservation, timber resistant to the penetration of preservatives.

Refusal point. (*proc*) The point at which no further significant amount of preservative can be forced into the timber by the method employed.

Relative humidity. (*phy*) The ratio, expressed as a percentage, of the actual mass of water vapour contained in a unit volume of air, to the mass of water vapour that would saturate that volume at the same temperature and pressure. Usually abbreviated to RH.

Remedial treatment. (*proc*) A treatment to timber permanently sited such as the constructional timber of a building where damage by wood-boring insects and/or wood-rotting fungi has already occurred.

Repellent. (*chem*) A substance or a physical condition which is unattractive or offensive to insects so that they move away from it.

Resinosis. (*for*) An abnormal exudation of resin or pitch from a conifer, also the abnormal impregnation of the woody tissue by resin. Of importance in wood preservation owing to associated differential absorption.

Resistant. (*chem*) Having the ability to withstand or impede. Used in two contexts in Wood Preservation. (A) Timbers may be said to be resistant to attack by specified wood-destroying organisms or (B) The extent to which a timber can be impregnated with preservatives are defined as:

Moderately resistant. Timbers fairly easy to treat. It is usually possible to obtain a lateral penetration of the order of $\frac{1}{4}$–$\frac{3}{4}$ in. in about 2–3 hours under pressure or the penetration of a large proportion of the vessels.

Resistant. Timbers difficult to impregnate under pressure requiring a long period of treatment. Often difficult to penetrate them laterally more than about $\frac{1}{8}$–$\frac{1}{4}$ in. Incising often used to obtain a better treatment.

Extremely resistant. These timbers absorb only a small amount of preservative even under long pressure treatments. They cannot be penetrated to an appreciable depth laterally and only to a very small extent longitudinally. See *Permeable.*

Reticulitermes. (*ent*) A genus of termites in the RHINOTERMITIDAE (dampwood and subterranean), including the following major pests of buildings, *R. flavipes* in North and Central America, *R. hesperus* in North America, *R. lucifugus* (and its variety or closely related species *santonensis*) in Europe, Middle East, North Africa and Madeira, and *R. speratus* in the Far East.

Retort. (*equip*) See *Pressure chamber.*

Retreatment. (*proc*) A term for the *in-situ* treatment of poles by bandaging.

Rhinotermitidae. (*ent*) A family of moist-wood subterranean termites, twenty-five species of which are serious pests of buildings. It contains the important genus *Reticulitermes* widespread in the north temperate zone roughly between latitudes 46 and 30°.

Rhizomorph. (*myc*) A compacted mass of hyphae of a fungus forming a cord-like structure. In the Dry Rot Fungus, *Merulius lacrymans*, the rhizomorphs convey water from moist to dryer wood.

Ridge. (*bldg*) The apex of a roof, generally a horizontal line.

Ridge board. (*bldg*) Sometimes known as ridge pole or ridge piece. The horizontal board, usually about 1 to $1\frac{1}{2}$ by 9 in. in cross section, set on edge, at which the rafters meet at their top.

Ridge tile. (*bldg*) A tile for covering a ridge. Four typical shapes are generally available in Britain. (1) Half-round, semi-circular of $7\frac{3}{4}$ in. internal diameter. (2) Hogback, which are half-round tiles and more acute at the apex. (3) Segmental, which are flattened half-round tiles, and (4) Angle tiles which have a sharp right-angle bend and flat surfaces on each side. Ridge tiles are usually 12 or 13 in. in length.

Rimu. (*for*) See *Dacrydium cupressinum.*

Ring-porous wood. (*bot*) Wood showing well-defined rings brought about by the pores of the spring wood being distinctly larger than those of the summer wood.

Riser. (*bldg*) The upright or vertical face of a step.

Rising-butt hinges. (*bldg*) Hinges which cause the door to rise about half an inch when opened. This is brought about by the helical bearing surfaces. The principal function is to clear carpets but additionally the door closes automatically and can be lifted off the hinges without unscrewing.

Rot. (*myc*) Deterioration or partial decomposition. In connection with wood, generally applied to the action of fungi and micro-organisms in which softening, loss of strength and mass, also a change in texture and colour, takes place. See *Brown rot*, *Dry rot*, *Heart rot*, *Soft rot*, *Wet rot and White rot.*

Round timber. (*for*) Felled trees, logs or poles before any longitudinal sawing operations.

Rueping process. (*proc*) An empty-cell process of preservative treatment in which compressed air is first applied to the timber. This is followed by the preservative liquid (often creosote) under a higher pressure which further compresses the air in the wood. On release of the hydraulic pressure, the air expands, forcing some of the preservative out of the wood to prevent bleeding. Patented in 1902 (Germany) by Wasserman, but operated by Rueping. See *Empty-cell process.*

Ruetgers' process. (*proc*) The Card Process in which creosote of high tar acid content was used. Generally in use between 1875 and 1885. Note that name of Ruetgers also associated with use of emulsifying agents in creosote/water/salt mixtures patented 1900 (J) and 1902 (G).

S

Salix. (*for*) A large genus of widely distributed hardwood trees generally known as willow. Osiers are the one year old shoots cut for basket-making. Cricket bats are made from *Salix coerulea* grown to about 18 in. diameter. Readily attacked by *Anobium punctatum*, especially old basketware. The sapwood is pressure impregnated with ease but the heartwood is resistant. Non-durable.

Sap. (*bot*) The fluid contents of plant tissue.

Sap displacement. (*proc*) Any preservative process in which the natural sap of green wood is replaced by a preservative liquid. The exchange is usually carried out by fastening a cap to one end of the bole of a freshly felled tree and either introducing aqueous preservative under pressure, or sucking out the sap by means of a vacuum. In the latter case, the other end of the tree stem is held in a container of preservative solution. The earliest form of sap displacement was achieved by leaving the branches on the tree and allowing transpiration by the leaves to pull preservative into the butt end and up the bole. The method is now applied with many variations and is particularly useful for treating timbers which, when seasoned, are unsuitable for pressure treatment. See *Boucherie process*, *Gewecke process*. See Plate 7.

Saprophyte. (*bot*) A plant which does not synthesise all its nutrient requirements from inorganic substances, but obtains them from dead organic matter.

Sapstain. (*myc, for*) Also called blue stain. A blue-black, blue-grey, brownish or purplish discoloration of timber, caused by fungal attack. Freshly felled softwoods are particularly susceptible and the fungi responsible belong, chiefly, to the genera *Ceratostomella* and *Hormodendrum*. In severe attacks the entire sapwood may be stained and although the appearance of the wood may be spoiled, thus reducing the market value, the strength of the wood is not necessarily impaired.

Not always confined to sapwood. Although most commonly occurring in coniferous timber (see *Blue stain*), it is also found in poplar, ash, and other hardwood species.

Sapstain prevention. (*proc*) As sapstain fungi, under suitable conditions, develop rapidly, preventive treatment must be carried out without delay, preferably within twenty-four hours of milling. Very rapid kiln-drying can prevent attack but as this is rarely convenient, a chemical dip of the freshly-milled timber is generally made where the problem occurs. Aqueous solutions of 2 per cent sodium pentachlorophenate or

1 per cent sodium pentachlorophenate with 2 per cent borax, have been widely used for this purpose.

Sapwood. (*for*) The outer layer of wood which, in the growing tree, contains living tissue and reserve food materials (eg starch). It is generally lighter in colour than the heartwood but not always clearly differentiated. The sapwood is generally much more susceptible to the action of wood-destroying organisms.

Sarking. (*bldg*) Used in Scotland only for roof boarding which may be up to $\frac{3}{4}$ in. in thickness.

Sarking felt. (*bldg*) Reinforced bituminous felt laid under the roofing tiles or slates. Roof boarding may or may not be present.

Sash. (*bldg*) The sliding light of a sash window.

Sash and case window. (*bldg*) In Scotland a sash window.

Sash cord. (*bldg*) A cord fixed to the side of a sash window which passes over a pulley and is held tight by a weight balancing the weight of the sash window, thus making it easy to open and close.

Sawn timber. (*proc*) Timber sawn to size but not having otherwise been machined.

Sclerotium. (*myc*) A resting body of variable size, composed of a hardened mass of hyphae.

Scolytidae. (*ent*) See *Scolytinae.*

Scolytinae. (*ent*) A sub-family of the CURCULIONIDAE, the Weevils, formerly a separate family, SCOLYTIDAE. They can be considered together with the PLATYPODINAE which were formerly the PLATYPODIDAE as both sub-families contain ambrosia beetles and bark beetles. The adults are usually heavily armoured and cylindrical, as are many wood-boring species in the ANOBIIDAE, LYCTIDAE and BOSTRYCHIDAE. The majority of species bore into the bark to the cambial layer where the larvae feed (bark beetles), whilst the larvae of others feed upon fungi which grow in the brood galleries (ambrosia beetles).

Seasoning. (*for*) The process of drying timber from the wet, freshly felled condition, to within the range of moisture content appropriate to the conditions and purposes for which it is to be used, by the method of air drying where the timber is exposed to more or less natural atmospheric conditions which may be followed by storage under more closely regulated conditions. Other methods of drying timber are high-temperature drying, kiln drying, solvent drying, vacuum drying, and vapour drying.

Septum. (*myc*) The cross-wall of a fungal hypha.

Seraya. (*for*) See *Shorea.*

Service records. (*bio/phy*) Collected data from actual installations of treated wood, which are used in comparing performance of different preservative treatments. These are principally kept by large-scale users of standard items such as poles and railway sleepers.

Service trial. (*myc*) A test on a limited scale, before large-scale use, of treated wood in the situations and circumstances in which it is required to be exposed.

Shingles. (*bldg*) Rectangular pieces of timber of shape and size similar to a roofing tile and used for the same purposes.

Shipworm. (*zoo*) Species of Mollusca which tunnel into wood floating or submerged in the sea such as eg *Teredo* species.

Shorea. (*for*) A very extensive genus of hardwood trees found chiefly in Malaysia, Indonesia, Borneo and the Philippines, the principal timbers of which are generally known as meranti, seraya and lauan, suitably qualified. Permeability to preservatives extremely variable even within one species.

Shothole. (*ent*) The gallery made by an ambrosia beetle (or shothole borer) not less than 1·5 mm and not more than 3 mm in diameter. It is associated with dark staining, does not contain boredust, penetrates timber mainly across the grain and is usually straight.

Shrinkage. (*phy*) A reduction in the dimensions of wood when water is lost from the cell-wall structure, often associated with unequal dimensional changes giving rise to twisting, bending, or to gaps between joinery timbers.

Siberian larch. (*for*) See *Larix*.

Siricidae. (*ent*) A primitive family of the Hymenoptera containing the woodwasps or horn-tails. It consists of large-sized, conspicuously-coloured insects, often a metallic blue or banded in black and yellow. The abdomen terminates in a short, triangular spine in the male, or is long and sharp in the case of the females. The ovipositor is long and sting-like and with it the insect drills holes through the bark of various trees and a single egg is laid in each tunnel in the outer sapwood. The species of economic importance attack softwood trees but certain hardwoods are attacked by species of a related but unimportant family, the XIPHYDRIIDAE. The larvae do considerable damage to timber in various parts of the world, and many species have become almost cosmopolitan pests, having been carried about the world in timber cargoes. An association with wood-rotting species of fungi has been shown. Two important species in Britain are *Urocerus gigas*, which is a dark-brown and yellow species, and *Sirex noctilio*, which is of metallic blue coloration.

Sitka spruce. (*for*) See *Picea*.

Skirting or skirting board. (*bldg*) A wooden board which may be decorated by a moulding at its top, fixed on edge to the foot of a wall. Its function is primarily for protection against kicks or marking by floor cleaning implements. In the USA it is known as washboard, scrub board or base, and in Scotland as the base plate.

Slab. (*proc*) The outer surface of a log removed during conversion so that it has one sawn, flat surface and one convex bark-covered surface.

Sleeper. (*proc*) A horizontal member on to which a rail is secured at right angles to it in order to distribute the load.

Slime mould. (*bio*) A member of the MYXOMYCETES, a primitive group related to, but not included in, the fungi. They are found on decaying wood as a rather amorphous but sometimes brightly coloured gelatinous mass and possess a number of characters usually associated with animals, such as glycogen used for food storage instead of starch.

Smoke generator. (*chem*) A finely divided insecticide incorporated into a pyrotechnic mixture which, when ignited, gives off a smoke composed of fine particles of the insecticide. A deposit of the insecticide is dispersed over a wide area but mainly on horizontal surfaces. Smoke generators are not fumigants since the insecticide is in the form of solid particles and not a gas.

Sodium fluoride. (*chem*) NaF. A fungicide and insecticide produced by the neutralisation of hydrofluoric acid produced by the action of acids on fluorspar or by the action of sodium carbonate on a mineral silicofluoride: see USP1,382,165.

It is a white, non-volatile powder of high melting point and density about 2·8. Its solubility in water at 18°C is 4·22 g/100 g. It is slightly soluble in ethanol.

Sodium fluoride is a powerful stomach insecticide with some activity by contact. It is highly phytotoxic as its use as a fungicide would lead one to expect. It was in wide use for many years as a sterilising fluid used in Dry Rot control. In addition, it was a component in the fluoride/arsenate/chromate/dinitrophenol waterborne wood preservative known as 'Tanalith U', 'Wolman Salts' or 'Osmosar' which was applied by pressure methods. 'Tanalith U' is covered by the BS3453:1962 which gives methods of estimation of fluoride content.

It is highly toxic to vertebrates, the lethal dose to man being 75 to 150 mg/kg, though it has been used as an ascaricide for pigs.

It was used as the commercial grade of 93 to 99 per cent purity but should be coloured as a warning of its poisonous properties.

Sodium pentachlorophenate. (*chem*) The sodium salt of pentachlorophenol which forms buff flakes with one mole of water of crystallisation; its solubility in water at 25°C is 33 g/100 g and it is insoluble in petroleum oils. The aqueous solution has an alkaline reaction and at solution strengths greater than 1 per cent causes skin irritation. The solid material irritates mucous membranes and may provoke violent sneezing. The technical materials are also known as 'Santobrite' (the Monsanto Chemical Co.) and 'Dowicide G' (the Dow Chemical Co.). The aqueous solution of sodium pentachlorophenate has been widely used as a sterilising fluid in Dry Rot outbreaks. Care must be taken to prevent the solution draining into ditches where it may ultimately be drained into fish-containing rivers, as it is particularly harmful to fish. See *Pentachlorophenol.*

Soffit or Soffite. (*bldg*) Generally the under-surface of building structure (except a ceiling or a floor), such as a stair, vault or a rib. Alternatively, the lining covering the opening produced by an overhanging element. Thus, a soffit board is a horizontal board fixed to the underside of rafters under an eave.

Soft rot. (*myc*) A decay of very wet wood caused by the action of a number of micro-organisms including species of bacteria and micro-fungi, but not usually by species of BASIDIOMYCETES. It commonly occurs in thin wooden cooling-tower slats, in mines, and also in wood floating or submerged in sea water. In the typical form of the decay there is a surface softening which is also darker in colour than normal wood. The softening gradually advances inwards. The growth rings stand out, as the early wood is attacked more severely than the denser late wood. Photomicrographs taken of sections of softwood across the grain show boreholes caused by the soft rot fungi which are usually near the middle layer of the cell wall. In freshwater, soft rot fungi are only important in alkaline conditions. Soft rots appear also to be generally more resistant to the action of wood preservatives than BASIDIOMYCETES.

Softwood. (*for*) The timber of trees in the class GYMNOSPERMAE, although almost all commercial timbers are confined to the order CONIFERAE or conifers. Characteristic of this latter group of trees is the possession of small, undivided, flat, or needle-shaped leaves. The leaves also exhibit water-conserving properties and with few exceptions (eg larch), are attached to the tree for several years ('evergreen').

Soil-block test. (*myc*) A technique for evaluating the fungicidal properties of wood preservatives. Screw top jars are used as containers and the test fungus is cultured on thin strips of untreated wood resting on damp soil. Soil is said to maintain a more even moisture content of the wood than in the agar-block test.

Soil poisoning. (*proc*) Now called 'soil treatment'.

Soil treatment. (*proc*) The application of insecticide, usually as a diluted water emulsion, to the area on which a building is to stand in order to inhibit the passage of termites from the soil to overlying wooden structures. This is generally carried out at the stage when foundations have been laid and all utility services have been connected, that is, when no further disturbance of the soil will take place.

Solute. (*chem*) Any substance dissolved in a solvent.

Solution. (*chem*) A solvent containing solute(s).

Solvent. (*chem*) A liquid capable of dissolving other substances. Usually the bulk of a wood preservative is a solvent which carries fungicides, insecticides, etc., into the wood, before (usually) itself evaporating.

Solvent-recovery process. (*proc*) The recovery of excess solvent from timber treated with an organic-solvent type preservative.

Sorption. (*phy*) The intake and retention of one substance (*the sorbate*) at the surface (*adsorption*) or in the interior (*absorption*) of another (*the sorbent*). Any subsequent loss of sorbate from the sorbent is termed desorption.

Southern white cedar. (*for*) See *Chamaecyparis*.

Southern yellow pine. (*for*) See *Pinus*. Includes timber of about six species. *Pinus palustris*, *Pinus elliotti*, *Pinus echinata*, *Pinus taeda*, *Pinus rigida* and *Pinus virginiana*.

Soxhlet. (*chem*) A distillation extraction apparatus used for testing the leachability of preserved wood. Water is boiled in a flask and the steam is led into a vertical double-surface condenser. The condensed water at 40 ± 5°C then drops into an extraction chamber in which the sample of preserved wood has been placed. When the extraction chamber fills it siphons back into the boiling flask and the cycle recommences. See *Leachability*.

Specific gravity. (*phy*) The weight of a given substance compared with the weight of an equal volume of water at the same, or at standard temperature and pressure. See *Density*.

Spore. (*myc*) In certain groups of plants, such as the Fungi, a one-celled structure which becomes separated from the parent plant to form the origin or starting-point of a new individual. A unicellular process of asexual reproduction. Spores are usually minute in size and often offer resistance to unfavourable conditions before the return of conditions suitable for germination.

Spore suspension. (*myc*) A suspension of fungal spores, sometimes including hypha fragments, in an aqueous medium such as water or normal saline. It is often used as an inoculum applied through an atomiser.

Sporophore. (*myc*) The spore-bearing or 'fruiting-body' of a fungus, the spores developing from cells in a special layer, the 'hymenium'.

Spray treatment. (*proc*) The application of preservative to timber by spraying. The liquid preservative is broken up into small droplets and directed by air pressure.

Spring wood. (*bot*) That part of the wood formed during the early stages of growth in each annual ring. It is characterised by the presence of large pores.

Spruce. (*for*) See *Picea*.

Square-edged. (*proc*) Sawn timber of rectangular cross-section throughout its length.

Stachybotrys atra. (*myc*) A species of fungus producing a mould, the spores of which are used in the preparation of a mixed spore suspension in the test for mould and mildew resistance of manufactured building materials. See BS1982:1968.

Stain. (*bio*, *chem*) Discoloration, sapstain and blue stain *q.v.*

Stair, stairs or staircase. (*bldg*) A series of steps of one or more flights with or without landings or larger horizontal areas where the direction may be changed, together with handrails and balustrades which give access from floor to floor (see BS565). Recommended minimum dimensions of house stairs are width 3 ft, treads $9\frac{1}{4}$ in., and going $8\frac{1}{2}$ in. A step of the stairs consists of one tread and one riser.

Standard. (*bldg*) Of timber, a measure consisting of 165 cu. ft (Petrograd Standard).

Standing timber. (*for*) Growing trees capable of conversion into timber.

Steam-and-quench treatment. (*proc*) A variation of the hot and cold open-tank treatment in which the timber is heated with steam before being immersed in the cold preservative. Sometimes combined with diffusion treatment.

Steaming. (*proc*) The application of live steam at atmospheric pressure or above to wood in a treatment chamber. This is usually carried out before treatment by the vacuum-pressure method to increase the permeability of the wood (pre-steaming). With some species, permeability of relatively green wood can be dramatically increased by so doing, although the moisture content is usually unchanged. It is believed that the heating action of steaming effects changes in those wood cell contents which normally restrict permeability.

Steeping. (*proc*) A preservative treatment in which the timber is completely immersed in a preservative solution usually for an hour or longer.

Sterigmata. (*myc*) The tiny protuberances on which the spores of basidiomycetes are borne.

Sterilise. (*myc*) To render incapable of reproduction or of conveying infection. To remove all unwanted living organisms from equipment or nutrient material. To kill any fungus present in a masonry wall by heating or chemical treatment or to kill insects in timber by heating in a kiln to a temperature not lower than 45°C and often as high as 60°C.

Stick(er). (*for*) A thin strip of wood 1×1 in., or $\frac{1}{2} \times \frac{3}{4}$ in., (25×25 mm or 13×19 mm) separating the courses (layers) when building up a stack of timber to allow unimpeded passage of air for seasoning or in pressure impregnation to permit circulation of the preservative.

Stipe. (*myc*) The stem bearing the cap or pileus of the sporophore in a number of basidiomycete fungi.

String or stringer. (*bldg*) A sloping board, two of which carry the treads and risers of a stair being cut or otherwise modified for the purpose. The inner is the wall string and the other is the outer string.

Structural timber. (*bldg*) Timber used in framing in buildings and in other load-bearing structures.

Sub-culture. (*myc*) A culture of a micro-organism made by taking an inoculum from a master culture and inoculating a sterilised nutrient medium. On the destruction of the master culture the new sub-culture becomes the master culture.

Sugi. (*for*) See *Cryptomeria japonica.*

Summer wood. (*bot*) That part of the wood formed during the later stages of growth in each annual ring. It is characterised by the presence of smaller pores. See *Spring wood.*

Surface-dry. (*for*) Applied to sawn timber, the surfaces of which are, to a greater or lesser extent, air dry.

Surface hardening. (*proc*) A condition of the surface of timber which resists the penetration of preservatives. It is caused by too rapid drying using heat. Even more imprecisely sometimes, by analogy with steel processing, referred to as case hardening.

Surface tension. (*phy*) The tension forces existing on the open surface of a liquid due to molecular attraction.

Surface timber. (*proc*) See *Dressed timber.*

Surface treatment. (*proc*) The application of a liquid preservative to the surface of timber. Ironically all wood preservative treatments are included under this heading, but by convention the treatments applied by brushing, spraying, and dipping are recognised as coming under this heading.

Surfaced timber. (*proc*) See *Dressed timber.*

Swelling. (*phy*) An increase in the dimensions of wood when water is taken up by the cell wall structure, often associated with unequal dimensional changes giving rise to twisting or bending.

Swellograph. (*phy*) A device giving a graphic record on a revolving drum of the increase in dimensions of a wood test sample, as a continuous function of time in contact with water or other fluids. See *Swellometer.*

Swellometer. (*phy*) A device consisting of a dial micrometer gauge linked with a vertical rod used for measuring the increase in dimensions of a wood test sample, as a function of time in contact with water or other fluids.

Swietenia. (*for*) A genus of hardwood trees from Central and South America, the West Indies and Cuba. The two principal species are *Swietenia macrophylla* and *Swietenia mahogani* and produce the timber known as mahogany, American, Spanish, Honduras, etc. 'True Mahogany'. A highclass timber for furniture and superior joinery. Rich golden brown in colour, but was usually stained red before french polishing. Resistant to attack indoors and resistant to fungal attack out-of-doors.

Sycamore. (*for*) See *Acer.*

Symbiont. (*bio*) One of the partners in a symbiosis.

Symbiosis. (*bio*) A relationship between two organisms of different species which is mutually beneficial, eg Termites and their gut-inhabiting Protozoa.

Synergism. (*chem*) The phenomenon of the enhancement of insecticidal effect of a substance by the addition of another substance (the synergist). Although insecticides are usually implied, there appears to be no reason why the synergistic effect should not also be applied to substances used for killing other groups of organisms, eg molluscicides, and fungicides. The syngergist may itself be ineffective at the concentration used in the mixture. Also referred to as potentiation or activation.

T

Tanalith C. (*name*) A proprietary brand of copper/chrome/arsenic wood preservative. See BS.4072:1966.

Tanalith U. (*name*) A proprietary brand of fluoride/arsenate/chromate/dinitrophenol water-borne wood preservative. See BS.3453:1962.

Tar oils. (*chem*) Tar oils are produced by the distillation of tars resulting from the high temperature carbonisation of coal and of coke-oven and blast-furnace tars. Although they have been used for wood preservation since 1890 the introduction of the formulated products known as tar oil washes or carbolineums for crop protection dates from about 1920. For these products the tar oils used are in the heavy creosote and anthracene oil ranges. These oils are brown to black liquids distilling from 230°C to the pitching point and of d^{15} 1·05 to 1·11. They are insoluble in water but soluble in organic solvents and in dimethyl sulphate. They consist mainly of aromatic hydrocarbons but contain components soluble in aqueous alkali: the 'phenols' or 'tar acids'; and nitrogenous bases soluble in dilute mineral acids: 'tar bases'.

Tarsus. (*ent*) The foot; the distal part of the leg of an insect consisting of from one to five segments and bearing the tarsal claws.

Teak. (*for*) See *Tectona grandis.*

Tectona grandis. (*for*) A hardwood tree from India, Burma, Thailand and Java, and introduced into many other tropical countries. The common name is Teak and this is the only true species. The tree is girdled some two or three years before it is felled in order that sufficient water is lost for subsequent flotation to take place. Teak, on account of its outstanding stability and durability, is one of the most valuable of all hardwoods. It possesses exceptional strength properties and is water and acid resistant, and the heartwood is highly resistant to fungal decay, termites and other insects. A natural oil is present in the wood giving it a somewhat greasy texture. An important property is the non-corrosion of metals with which it comes into contact including nails and screws. The low degree of fire resistance which is claimed for teak stems from the fact that teak ignites readily and flame spreads quickly charring all surfaces evenly rather than burning through at, say, bottom of door or bulkhead. It is used for a wide variety of purposes including high-class joinery fittings and bench tops in laboratories and chemical works, bridges, railway construction and carriages, shipbuilding and an extensive range of outdoor usage.

Tension wood. (*bot*) Modified woody tissue found in the timber converted from broad-leaved trees which have been grown in a leaning position. It occurs on the upper, uphill, side of the tree and is characterised by being lighter in colour and more lustrous than normal wood. It is exceptionally weak in compression parallel to the grain and like compression wood has abnormally high longitudinal shrinkage on drying.

Teredo. (*zoo*) A genus of bivalve molluscs the species of which tunnel into wood floating or submerged in the sea. *Teredo*-attacked timber is identified by the calcareous coating to the galleries, the minute hole communicating with the sea-water and the complete discreet nature of each tunnel. The adult *Teredo* never leaves the timber in which it is virtually imprisoned but the reproductive elements are voided from the delicate siphons which project from the communication aperture See Plate 10.

Termites. (*ent*) See *Isoptera.*

Termite shield. (*proc*) A mechanical device usually of metal fastened to the vertical supports of buildings in order to prevent the overbuilding of termite tubes and, thus, to minimise the access of subterranean termites.

Termiticide. (*chem*) A substance which kills termites (Isoptera).

Termitidae. (*ent*) A family of ground-dwelling, mound-building and tree-nesting termites of which seventeen species are serious pests of buildings. Species of *Macrotermes* and *Odontotermes* are found in Africa, India and south-east Asia, and of *Nasutitermes* in Ceylon, Mauritius, Australia, Central America and the West Indies.

Thorax. (*ent*) The second, or middle, main division of the body of an adult insect.

Through-and-through sawing. (*for*) The conversion of logs by a series of lengthwise parallel cuts.

Thuja. (*for*) The timber of *Thuja plicata* from N.W. America commonly known as Western Red Cedar. At first it is reddish-brown but weathers to grey and contains aromatic oils imparting some resistance to wood-destroying insect attack. See *Cedrus.*

Tie. (*bldg*) The word applied in Scotland to a clip for fixing flexible metal roofing sheets.

Tie-beam. (*bldg*) The horizontal, lowest, member of a roof truss of the same length as the span. It is fixed to the feet of the rafters of opposite sides by means of a heel strap.

Tilia. (*for*) A genus of north temperate hardwood trees. *Tilia vulgaris* and two other species are known as European Lime. *Tilia japonica* and other species are known as Japanese Lime; *Tilia americana* and some other species are known as basswood, from eastern Canada and USA. The timber is white, straight-grained, soft, and inclined to woolliness. One of the most important timbers used for carving, particularly silk screen blocks and the like, but has been used for many other purposes, including musical instrument construction.

Timber. (*for*) Wood in a form suitable for construction, carpentry, joinery or manufacture. Growing, standing, or felled trees capable of being converted to these uses.

Timber Research and Development Association or TRADA. (*name*) When first formed in 1934 this was known as the Timber Development Association—TDA. During World War II effort was mainly directed towards the Defence Departments in connection with aircraft, boat and hutting construction. Then, the Association played an important part in the identification and testing of the many new commercial timbers which were at the time becoming available from overseas sources. After the war TDA moved into the building research field and proposed new design solutions for domestic roofs, glued laminated timber construction and large timber engineered structures—substantial achievements which enabled wood and wood-based products to gain prominence in the construction industry. In 1955 TDA made its first formal links with the DSIR (now Ministry of Technology) and by 1962 it had become a research association with Government support. In 1963 the title was changed to Timber Research and Development Association.

The number of publications available from TRADA at 1st August 1968 was about 370 titles covering a wide field of interests and subjects. Major publications published in recent years are (1) *The Design Guide to Timber Frame Housing* which, when completed, will comprise 250 data and design sheets, (2) *The Final Report on the TRADA Timber Handling Investigation* (292 pages illustrated with graphs, diagrams, and drawings), (3) *TRADA Wall Charts on Wood Technology* (a series of eighteen coloured, large-size wall charts covering all aspects of timber processing, manufacture and utilisation), (4) *Explanatory Memoranda on Building Regulations* (a series of three covering England and Wales, Scotland and the GLC area).

Address: Hughenden Valley, High Wycombe, Buckinghamshire,
and The Building Centre, Store Street, London, WC1.

Topical application. (*tox*) A method of administration of a substance by placing it directly upon the surface or integument of the test subject. Widely used in assessment of insecticides by the precise application of a measured amount of substance through the use of a carefully calibrated micrometer syringe, and a holding device.

Total inhibition point (TIP). (*tox*) The lowest concentration of a particular preservative that will prevent the development of a particular fungus or insect, under specified conditions, without necessarily causing its death.

Toxic hazard. (*tox*) The extent to which man, or other organisms, are exposed to harmful effects due to the use of a specified substance.

Toxicity. (*tox*) A measure of the harmful effect produced by a substance on a living organism.

Toxic limit. (*tox*) The interval between that concentration of the preservative which just permits decay due to a specified fungus and the concentration next highest in the series which inhibits all decay under the test conditions, or in the case of a specified insect that concentration of the preservative which permits survival of the larvae and the concentration next highest in the series which kills them.

Toxic threshold. (*tox*) Toxic limit.

Tracheid. (*bot*) A needle-like vascular cell with pitted lignified walls serving for the conduction of water and for imparting mechanical rigidity. The characteristic elements of softwood tissue which when viewed in transverse section appear tightly packed together in a honeycomb-like pattern. Some tracheid-like cells, however, are found in certain hardwoods such as oak and chestnut.

Trametes serialis. (*myc*) A basidiomycete fungus commonly decaying coniferous logs causing a brown, crumbling rot similar to *Poria monticola*. Tough corky sporophore is cream or buff and irregular when found in a building. Pores measure 1–8 mm in length.

Trametes suaveolens. (*myc*) A basidiomycete fungus associated with Death Watch Beetle attack.

Trap tree. (*proc*) A dead or girdled standing tree left in the forest to attract wood-boring insects and so facilitate their destruction or to maintain a reservoir of infestation in the likely event of the trap tree not being dealt with at the correct time.

Tread. (*bldg*) The horizontal or level part of a step on which the foot is placed.

Tree injection. (*proc*) The introduction of a solution of a water soluble salt into the sap stream of a living tree, often under pressure, in order to kill it, to protect it from disease, to kill parasitic plants such as mistletoe, or into a freshly killed tree in order to control wood-boring insects.

Tributyltin oxide. (*chem*) TBTO, Bis (tri-n-butyltin) oxide. A colourless or light yellow liquid of molecular weight 596.

Boiling point 180°C at 2 mm Hg. Freezing point below −45°C. Specific gravity (d_4^{20}) 1·14. Flash point—above 240°F (Pensky Martens). Viscosity 4·8 centistokes at 25°C. This substance has enjoyed a wide use as a fungicide and bactericide in paint, paper, and leather manufacture and finds an increasing application in wood preservation where a wide spectrum of fungicidal activity is required. Commercial development of organotin compounds is relatively new. In 1942 they were used to stabilise polyvinyl-chloride but the impetus to their use is due to the work of G. J. M. van der Kerk in 1949.

$$(n\text{-}C_4H_9)_3C\text{-}Sn\text{-}O\text{-}Sn\text{-}C(n\text{-}C_4H_9)_3$$

The oral mammalian toxicity of TBTO is considerable, the oral LD_{50} for rats having been given as 194 mg/kg, although the concentration required in commercial wood preservatives is low. T. Hof and J. G. A. Luijten have given the following percentage toxic limits for some wood-destroying fungi as determined by the agar test.

Polyporus vaporarius	0·0001—0·0002
Polystictus versicolor	0·0008—0·0016
Coniophora puteana	0·0008—0·0016

Trimmed joist. (*bldg*) A common joist which has been cut short at an opening and is fixed to a trimmer joist.

Trimmer joist. (*bldg*) A short joist which encloses one side of a rectangular hole in a wooden floor. It is usually matched by another to form the side opposite and parallel to it.

Trimming joist. (*bldg*) A joist parallel to the common joists and fixed each side of a hole in a wooden floor. The trimmer joists make up the other two sides.

Triplochiton scleroxylon. (*for*) A hardwood tree species from West Africa giving a very light white timber known as Obeche or Wawa if originating from the Gold Coast. Subject to blue-stain, pin-hole borers, non-durable. Uses: interior joinery fittings, core veneer in plywood.

Truewood. (*for*) The central core of wood which in the growing tree has ceased to contain living cells. Generally darker in colour than the outer sapwood, though not always clearly differentiated from it. Generally used in connection with Australian hardwoods.

Tylose. (*bot*) An intrusion of parenchymatous tissue into a vessel of a hardwood through a pore. The resistance to penetration of preservatives and to the spread of fungal hyphae in certain hardwoods is probably due to the presence of tyloses.

Tyre-tube process. (*proc*) A variant of the boucherie process in which small, green, unbarked timber such as fence posts, has water soluble salt solutions applied at the butt ends by hydrostatic pressure by the fitment of sections of tyre tubes.

U

Ulmus. (*for*) A genus of north temperate hardwood trees generally known as elm. Their standard names are *Ulmus americana* from eastern Canada and USA, White Elm; *Ulmus glabra* from the British Isles, Wych Elm; *Ulmus hollandica* from the British Isles and Europe, Dutch Elm; *Ulmus carpinifolia* from Europe including British Isles, Smooth-leaved Elm; *Ulmus procera* from the British Isles, English Elm; *Ulmus thomasii* from eastern Canada and USA, Rock Elm and various *Ulmus* species but chiefly *Ulmus lacinista* and *Ulmus davidiana*, Japanese Elm. The timber is generally tough and difficult to split and its rather twisting grain causes it to warp during seasoning if insufficient care is taken. It is usually not considered to be durable, but if completely submerged such as in piling or bridge foundations it is one of the most successful of timbers. It is also widely used for furniture, weatherboarding and wheelbarrows.

Unseasoned. (*for*) Freshly felled timber still containing liquid sap.

V

Vacuum. (*phy/proc*) Any gas pressure below atmospheric, produced by some kind of pressure-reducing pump. The ultimate vacuum that can be achieved is a completely gas-free void. Very high vacuums are often expressed as the height of mercury column that would be supported against a complete vacuum. The unit of measurement is the Torr (=1 mm Hg). In wood preservation, such high vacuums are rarely reached and the most convenient way of expressing the degree of vacuum is as a percentage of the ultimate achievable. Values of 90–95 per cent are typical in full-cell vacuum pressure treatment whilst about 40 per cent is more normal in double-vacuum treatment.

If a vacuum is applied to dry timber before the introduction of preservative liquid, it is called a 'dry vacuum'. Conversely, a 'wet vacuum' involves nearly filling the treatment chamber with preservative before pumping the remaining gas-space down to the full degree of vacuum required. It appears that good penetration of a particular wood species, especially when not fully seasoned, can depend on whether a wet or dry vacuum is used. Either may be more suitable in a particular instance.

Vacuum chest. (*proc*) A tank installed between the vacuum pump and the treatment chamber of a preservation plant which prevents preservative from entering the pump inlet.

Vacuum desiccator. (*bio*) A piece of laboratory apparatus designed to maintain specimens in a dry condition by means of the provision of a lower chamber containing a strongly hydroscopic material such as strong sulphuric acid or calcium chloride, and also fitted with a stopcock by means of which connection may be made to a vacuum pump.

Vacuum pressure process. (*proc*) Any process for the treatment of timber in a pressure cylinder, in which the wood is first subjected to a vacuum before injection of the preservative under pressure.

Vacuum process. (*proc*) The application of a preservative to timber in a treatment cylinder by evacuation of the air, then filling the cylinder with preservative, and finally releasing the vacuum. Generally used for treatment of permeable timbers with preservatives of low viscosity.

Valley. (*bldg*) The intersection at the lowest level of two inwardly sloping surfaces of a roof towards which water flows.

Valley board. (*bldg*) A wide 1 in. board fixed on to and parallel to the valley rafter.

Valley gutter. (*bldg*) A gutter in a valley usually lined with lead or other flexible metal in older construction, but can be of concrete precast or cast in place.

Valley rafter. (*bldg*) A rafter carrying a valley gutter or a valley board.

Vanillin phloroglucinol. (*chem*) A test used for detection of organochlorine based wood preservatives in wood. It is based on the pyrolysis of the organochlorine compound and detection of the generated hydrochloric acid gas. For details of test see draft BWPA Standard 'Pentachlorophenol with Insecticides'.

Veneer. (*proc*) A thin ply of wood. The plies are usually glued together in layers with the grain alternating to form plywood. A veneer of high quality timber or of special decorative value is often applied on an outer face.

Vessel. (*bot*) An axial series of cells which have joined together to form a tube-like structure of indeterminate length serving for conduction and mechanical strength. The walls are generally furnished with pits. A characteristic element of hardwoods.

Vessel porous. (*bot*) A classification of hardwoods based on the presence of very long continuous vessels with few barriers such as end-plates or tyloses. Such timbers are very permeable to fluids in the longitudinal direction, and rapid penetration of preservative can be attained along the length of a piece. Despite this, it does not follow that the fibres and other wood cells necessarily receive an adequate retention to prevent decay and a more intensive treatment may be necessary.

Virginian pencil cedar. (*for*) See *Juniperus*.

Volume/volume. (*chem*) The ratio of the volume occupied by one component in a mixture to the volume of the mixture as a whole. Often expressed as a percentage. See *Weight/weight*.

W

Wainscot. (*bldg*) Panelling, generally of wood, up to dado height. Sometimes called wainscoting.

Wall-boards. (*bldg*) Sheets of rigid manufactured material for covering large areas. They are made from a wide variety of materials, including several kinds of wood fibre and plywood.

Wall plate. (*bldg*) A timber usually of substantial dimensions fixed along the top of a wall at eaves level and on to which bear the joists and/or rafters.

Walnut. (*for*) See *Juglans.*

Wane. (*for*) The presence of the original sapwood surface, with or without bark, on a face or edge of square-sawn timber.

Waney edge. (*for*) See Wane. The presence of the bark on softwood timber is necessary for an infestation of *Ernobius mollis* to occur.

Warp(ing). (*for*) Distortion in converted timber causing twisting, etc., usually developing during seasoning.

Water-borne preservative. (*chem*) A wood preservative in the form of a solution in water.

Watermark. (*for*) A patterned stain in certain timbers due to bacterial or fungal attack in the living tree. In some cases the value of the wood for decorative purposes is increased.

Water repellent preservative. (*chem*) A wood preservative formulated to give timber to which it is applied, the property of retarding the rate of absorption of liquid water.

Wawa. (*for*) See *Triplochiton scleroxylon.*

Weathering. (*phy*) The effect of climatic conditions such as rainfall, relative humidity, sunlight and wind, on wood sited externally. Checking and splitting of wood caused by climatic conditions.

Weathering cycle. (*phy*) An attempt in the laboratory to reproduce the changing pattern of rainfall and sometimes other climatic phenomena, at regular intervals, on a wood test sample.

Weatherometer. (*phy*) A device used in the laboratory to reproduce an accelerated effect simulating the changing pattern of rainfall and sometimes other climatic phenomena on wood test samples, at regular intervals.

Weevils. (*ent*) See *Curculionidae.*

Weight/volume. (*chem*) The concentration of one component in a solution, usually expressed as the weight of the component in grams contained in 100 ml of solution.

Weight/weight. (*chem*) The ratio of the weight occupied by one component in a mixture to the weight of the mixture as a whole. Often expressed as a percentage. See *Volume/volume.*

Western red cedar. (*for*) See *Thuja.*

Western larch. (*for*) See *Larix.*

Western white spruce. (*for*) See *Picea.*

Wet rot. (*myc*) See *Coniophora cerebella.*

Wharf-borer. (*ent*) See *Nacerdes melanura.*

White ant. (*ent*) Used colloquially, but now becoming obsolete, for termite, for which see *Isoptera.*

White pocket rot. (*myc*) A decay, usually of standing timber, characterised by the presence of a number of white, oval to circular pockets from which the cellulose has been digested by the fungus. *Stereum padiceum* is the cause of a white pocket rot in oak.

White rot. (*myc*) A decay of wood caused by a fungus which has decomposed all the wood components including the lignin. The term 'corrosion rot' has also been proposed. The wood turns white or very light in colour, and frequently the zones of advanced decay are outlined by a dark-coloured band. A general thinning of the cell walls of the wood is a feature of an advanced stage of the decay. Examples of fungi causing a white rot of wood are *Polystictus versicolor* and *Phellinus cryptarum.*

Whitewood, American. (*for*) See *Liriodendron tulipifera.*

Wild grain. (*for*) Irregular grain running in various directions, often within a restricted area. This may cause irregular absorption of wood preservatives.

Willow. (*for*) See *Salix.*

Winder. (*bldg*) A triangular or wedge-shaped tread of a stair which thus changes its direction.

Wolman process. (*proc*) A double impregnation process using creosote oil and a solution of triolith (a mixture of fluorides, nitrophenols and chromates).

Wolman salts. (*proc*) A generic name for a number of water-soluble preservative salts developed and marketed by Dr Wolman GmbH, whose antecedent company, Oberschlesiche Holzimpraegnierung GmbH, had been founded in 1903. Sodium fluoride was used in 1907, followed in 1913 by Triolith, a mixture of fluorides, nitrophenols and chromates. Then in 1921 the addition of arsenic compounds gave a mixture known as Tanalith. In 1930 the chromate content was increased and the products were called Triolith U and Tanalith U. In Germany, these are today marketed as Wolmanit U and Wolmanit UA, and Basilit U and Basilit UA. All are generally known as Wolman Salts.

Wood block. (*myc and ent*) A test block of wood carefully selected for characteristics of standardisation and of non-durable species such as *Pinus sylvestris.* In BS838:1961 Methods of Test for Toxicity of Wood Preservatives to Fungi, the dimensions of the wood blocks are $5 \times 2{\cdot}5 \times 1{\cdot}5$ cm and the long axis is parallel to the grain of the wood.

Wood plastic composite (WPC). (*proc/chem*) Wood in which some or all of the air spaces have been filled with a plastic substance. This is usually achieved by impregnating the wood with a monomeric fluid, either by simple immersion or by the use of pressure, followed by a solidifying polymerisation stage. Polymerisation can be initiated either chemically, by the use of heat or by irradiation with gamma rays. Wood plastic composites usually have greater moisture resistance, strength, and particularly hardness, than the untreated wood.

Wood pulp. (*proc*) Wood reduced to pulp for the manufacture of paper and fibre-board. This is carried out either by grinding and washing or by the removal of lignin and the chemical maceration of the woody tissues with sulphite or soda solutions. Decay of the wood before pulping, and deterioration of the manufactured pulp due to fungal attack, has often been a problem.

Wood substance. (*chem*) The chemical substances of which the cell walls of wood tissue are composed.

Wood wasp. (*ent*) See *Siricidae*.

Wood wool. (*bldg*) Thin, narrow shavings of constant dimensions but of indeterminate length made from light straight-grained wood free from odour, resin, and gum, and used for packing; also as a component in a number of building structures in order to give lightness.

Woodworm. (*ent*) An imprecise expression for the larvae of insects which bore in wood. In Britain has come to signify the larvae, particularly of *Anobium punctatum* (sometimes referred to as 'Common Woodworm'), but also of *Lyctus* and Death Watch Beetle which damage wood utilised by man. It should be noted, however, that the *Oxford English Dictionary* includes *Teredo* or Shipworm within its definition, going back to 1540 for its authority.

Wormhole. (*ent, bio*) A hole or tunnel caused by a wood-boring insect or marine borer. In sawn timber a round hole in transverse section may denote the tunnel of an ambrosia beetle or a wood wasp larva. Alternatively the exit hole of a species of ANOBIIDAE, LYCTIDAE, CERAMBYCIDAE, BOSTRYCHIDAE, etc.

Wormy. (*for, ent*) The condition of timber which has been attacked by woodboring insects either when freshly felled (SCOLYTIDAE and PLATYPODIDAE) or when converted and utilised (CERAMBYCIDAE, ANOBIIDAE, LYCTIDAE, etc.).

X

Xestobium rufovillosum. (*ent*) Death Watch Beetle in the family ANOBIIDAE which occurs particularly in southern England and the Midlands. It is found attacking only the hardwood timbers in older buildings and is generally associated with fungal decay. It is the largest of the British ANOBIIDAE being from 5 to 7 mm in length and is dark chocolate brown in colour, with a number of patches of yellowish scales which rapidly rub off. It usually emerges at the end of April and about forty to sixty eggs are laid. The young larvae, unlike those of *Anobium punctatum*, are agile and crawl over the wood surface before entering the wood at a crevice. The relatively large bun-shaped faecal pellets are of characteristic shape. The length of the larval stage is extremely variable according to the degree of fungal decay which has taken place. This was found experimentally to range from ten months to over ten years. The larvae pupate in late summer or early autumn and the pupal stage lasts three or four weeks, but the adult beetle remains within the pupal chamber throughout the winter and early spring. They are rarely seen to fly in nature and the length of adult life is about ten weeks. The mating call or tapping of the Death Watch is a well-known phenomenon and is brought about by a series of rapid blows of the head against the wooden surface on which it is sitting. It is found throughout Europe, in North Africa, and has been found in a few scattered localities in the USA where it has obviously been introduced.

X-rays. (*phy*) Röntgen rays. Electromagnetic waves of a similar type to light but of much smaller wavelength, in the range of 5×10^{-7} to 6×10^{-10} cm. Produced when cathode rays strike a material object. The absorption of the rays by matter depends upon its density and atomic weight. Soft X-rays up to 30 kv produced by equipment with a beryllium window are used for examining blocks in Larval Transfer tests. See BS3651:1963.

Xylaria hypoxylon. (*myc*) A basidiomycete fungus associated with Death Watch Beetle attack.

Xylem. (*bot*) The tissue of stem and root of a woody plant lying between the cambium on the outside and the pith in the centre. Generally the majority of the cells composing it are dead, but the living cells of the xylem serve for food storage and its translocation as well as water conduction. The dead cells serve to give mechanical strength. Xylem tissue (more specifically secondary xylem) is the timber of commerce.

Xylocopidae. (*ent*) Carpenter bees. A number of species in this hymenopterous family bore into sound wood in several parts of the world. *Xylocopa orpifex* is found throughout the eastern states of North America, whilst *X. virginica* is found in the west. The borings made by this latter species are about ½ in. in diameter and from 5 to 18 in. in length. Other species in East Africa and elsewhere necessitate remedial treatment.

Xylophaga. (*zoo*) A genus of Mollusca related to *Teredo*, the species of which tunnel into wood submerged or floating in the sea. See *Marine borers*.

Y

Yellow cedar. (*for*) See *Chamaecyparis*.

Z

Zinc chloride. (*chem*) $ZnCl_2$ A white crystalline salt, very soluble in water. This was one of the earliest inorganic water-borne preservatives and fire retardants used commercially. Because of its water solubility, zinc chloride was not very successful for outdoor applications, and it was very corrosive towards treatment plant metals. Its use has long been discontinued.

Zone line. (*for*) A black or coloured line formed in wood in the process of decay by certain wood-destroying fungi often delimiting small irregular areas.

BIBLIOGRAPHY

ANON, 1955, *Nomenclature of Commercial Timbers including Sources of Supply.* BS881 and 589:1955. British Standards Institution, London.

ANON, 1956, *A Handbook of Hardwoods.* Forest Products Research. HMSO, London.

ANON, 1963, *Glossary of Terms relating to Timber and Woodwork.* BS565:1963. British Standards Institution, London.

CORKHILL, T., 1948, *Glossary of Wood.* Nema Press, London.

FLOOD, W. E., 1960, *Scientific Words, Their Structure and Meaning.* Oldbourne, London.

IMMS, A. D., 1957 (Ninth edition revised by Richards, O. W. and Davies, R. G.), *A General Textbook of Entomology.* Methuen, London.

KENNETH, J. H., 1967, *Dictionary of Biological Terms,* 8th Edition. 1st to 7th Editions by Henderson, I. F. and Henderson, W. D. Oliver & Boyd, Edinburgh.

MARTIN, H. (Ed.), 1968, *Pesticide Manual.* British Crop Protection Council.

SCOTT, J. S., 1958, *A Dictionary of Civil Engineering.* Penguin Books, Harmondsworth, England.

SCOTT, J. S., 1964, *A Dictionary of Building.* Penguin Books, Harmondsworth, England.

UVAROV, E. B. and CHAPMAN, D. R., 1942, *A Dictionary of Science.* Penguin Books, Harmondsworth, England.

WATERHOUSE, D. F. (Ed.), 1970, *The Insects of Australia.* CSIRO, Melbourne University Press.